# CALCULUS

## WITH THE HEWLETT PACKARD GRAPHING CALCULATOR

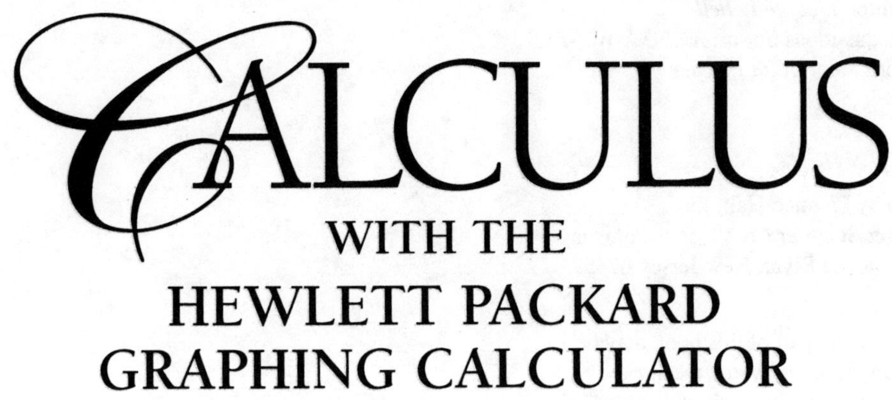


# CALCULUS WITH THE HEWLETT PACKARD GRAPHING CALCULATOR

JACK K. COHEN ▪ FRANK G. HAGIN

Colorado School of Mines

PRENTICE HALL, Upper Saddle River, NJ 07458

Production Editor:*Ann Marie Longobardo*
Production Supervisor: *Joan Eurell*
Acquisitions Editor: *George Lobell*
Supplements Acquisitions Editor: *Audra J. Walsh*
Production Coordinator: *Alan Fischer*

Simon & Schuster / A Viacom Company
Upper Saddle River, New Jersey 07458



ISBN: 0-13-520339-2

Printed in the United States of America

10 9 8 7 6 5 4 3 2 1

Prentice-Hall International (UK) Limited, *London*
Prentice-Hall of Australia Pty. Limited, *Sydney*
Prentice-Hall Canada Inc., *Toronto*
Prentice-Hall Hispanoamericana, S.A., *Mexico*
Prentice-Hall of India Private Limited, *New Delhi*
Prentice-Hall of Japan, Inc., *Tokyo*
Simon & Schuster Asia Pte. Ltd., *Singapore*
Editora Prentice-Hall do Brasil, Ltda., *Rio de Janeiro*

# Contents

## REMARKS TO STUDENTS

A modern computing device, like the HP-48, is an important tool for working scientists. The exact tool that you will use when you begin your career will no doubt be different, probably much more powerful; but it makes sense to begin developing skills in using this remarkable technology. Remember, however, that your primary task is to learn calculus, not be become a whiz on this or some other computing device. Thus, the projects in this manual focus heavily on a small group of commands and programs and only hint at more advanced usage.

We have tried to ease the burden of mastering the HP-48 by providing numerous examples, warnings about common pitfalls and, when appropriate, supplying programs which you can use and modify as needed. But this is not a replacement for the User's Guide.

You may want to go more deeply than we will into the many graphing options offered by the HP; and you may want to become more skilled in programming than is required for the simple programs given in this manual. You will no doubt find a growing number of opportunities to use your programmable calculator in other courses.

The projects in this booklet are viewed as "weekly" in that they are, for most calculus texts, spaced at about one week intervals for three semesters. Most projects consist of three parts:

**Before the Lab.** Relevant reading and warm-up problems. These problems are important to the course as well as to the project.

**The Lab.** Problems requiring the HP. Your instructor will tell you which problems to do. This section will typically take about an hour of lab work.

**After the Lab.** You'll be asked to reflect on what your computations revealed.

Your instructor may well shorten some of the projects by telling you to omit certain exercises. We suggest that you read over the omitted exercises and give brief thought as to how you would tackle them. You will also note that we have labeled certain projects as "optional." This designation is given to projects that either use advanced HP facilities that won't be needed in the sequel or that involve a greater degree of exploration on your part. Again, even if your instructor omits some or all of these projects because of time pressure, we suggest that you read them—you may well want to refer back to these projects later in your academic or scientific career.

As mentioned below in the *Remarks to Instructors*, the projects were originally written for a powerful software package run on computers. In making the HP version of the booklet we had to omit a few projects that were not suited for your device; and you will not see those projects.

We hope that these projects will add to your calculus course—both by increasing your knowledge and by increasing your interest. Your Authors welcome your comments for improving this manual. Our Internet addresses are given below at the end of this Introduction.

## REMARKS TO INSTRUCTORS

For the past three years at the Colorado School of Mines, all 500-plus freshmen have taken a calculus course using major portions of the material you have in your hands. The authors each have over 25 years of teaching experience at all university levels from freshman to Ph. D. level graduate courses. We have participated fully in all aspects of university life and are well published applied mathematicians. However, our recent experience with this calculus program ranks among the most exciting and rewarding of our careers—it added a distinct element of fun compared with our previous calculus experiences. However, while "fun" is not to be underestimated as a motivating force, we have also kept firmly in mind a number of more concrete pedagogical objectives. We explain our viewpoint in the following paragraphs.

Our guiding star is that we are teaching calculus, not computing technology. This has important implications. For example, to our mind, premature reliance on built-in HP capabilites (e.g. `SOLVE`, for finding the roots of equations) can easily degenerate the computational experience to being the high-tech analog of looking up the answer in the back of the book. These "black box" tools often do not help teach calculus, so we relegate them to a secondary role. For example, in the case of `SOLVE`, we introduce it only after presenting explicit coding for Newton's method. The instructor notes to such projects will clearly state these projects are not of primary importance to learning calculus and may be skipped. In particular, succeeding projects will *not* depend on covering these "black box" projects. Consider at least making some reading assignments in some of these projects, but don't let them deflect you from your main task of teaching calculus.

To minimize potential frustrations and the time involved, we often provide the required command structures or simple versions of programs so that students (and faculty) can copy and do minor editing.

We have firm objectives in teaching calculus and these projects are designed to reinforce them:

**Solve general problems.** Instead of solving problems with numbers that make the answers "come out nice," solve *families* of problems. That is, we emphasize the use of parameters for as long as possible, putting in specific numerical values only when necessary for the final calculation.

**Emphasize fundamentals.** Don't rush through hundreds of topics, instead take time with the topics that deserve it, and train students to *use* their text, when necessary, to look up topics that are less important.

**Empower students.** The picture of two or three students huddled around a computer or calculator is often a better image than 35 students busily taking notes in a lecture hall. We allow our students to hand in joint work for all the projects—and we encourage you to think about doing this too.

As stated above, most projects consist of three parts:

**Before the Lab**, **The Lab** and **After the Lab**. The intent of this division is to make the Laboratory experience more meaningful.

For each project, the Instructor's Answer Booklet contains:

a) Answers (naturally!).

b) An introductory statement of the goals of the project.

c) Suggestions for ways to shorten the project (you'll probably want to vary the assignments from semester to semester or section to section).

d) Occasional Instructor Notes about likely student questions, etc.

We have been generous in both the number of projects and their contents. For most projects, we list exercises that can be omitted to yield a shorter project. We have also labeled certain projects as "optional." This designation is given to projects that either use "black box" HP commands that won't be needed in the sequel or that introduce modest "discovery" explorations. These projects can be among the most exciting, so we hope you won't omit *all* the optional projects. Similarly, we do not expect that you will assign *all* the regular projects.

A final note. The original projects developed at CSM were written for the powerful *Mathematics* package in a well-equipped NeXT laboratory environment. Upon entering our publication agreement with Prentice-Hall we were convinced to make versions for the *Maple* and *MATLAB* packages, and for the TI-85 and HP-48 calculators. Frankly, we were not optimistic that the majority of these projects were suitable for the hand-held calculators, as impressive as this technology is. We were both surprised and gratified that the large majority of projects were *very* well suited for these remarkable little devices. It makes one look forward with excitement to the next generation of the programmable calculator!

## ACKNOWLEDGMENTS

Over these past three years, we have written numerous worksheets and projects containing problems to be solved using computing technology. Most of them have significant original content, but many have been motivated by or use material from articles written in teaching-oriented journals. In particular, many of our project questions originated from routine textbook exercises. Similar exercises could be found in many calculus texts, but since we used the popular Edwards and Penney Third Edition during the past three years, often the original routine kernel exercise was drawn from this excellent text and we wish to acknowledge that.

We owe a great debt to our colleagues on the "Calculus Team" at the Colorado School of Mines, who willingly tested earlier versions of these projects. In particular, the advice and encouragement of the team coordinator, Barbara Bath, was invaluable. We also wish to acknowledge our grader, Shelby Worley, for his astute comments about student difficulties with the earlier versions of these projects. Our questions are distinctly more "student friendly" because of his efforts. Finally, we give our thanks to the students of the School of Mines for their enthusiasm and advice.

## ADDRESSES

We welcome your comments for improving this manual.

Jack K. Cohen
Center for Wave Phenomena
Colorado School of Mines
Golden, CO 80401
jkc@keller.mines.edu

Frank G. Hagin
Department of Mathematics
Colorado School of Mines
Golden, CO 80401
fhagin@newton.mines.edu

## BEFORE THE LAB

**Beginning with your HP**: We shall provide some guidance in using the HP, but we cannot reproduce the User's Guide in this manual. We expect you to look up things as needed or just to experiment (people differ in their tolerance for just "playing around" until they figure out how things work, versus their willingness to search through manuals).

**Keystroke Conventions**: We denote the HP violet leftshift key by [⇐] and the aqua rightshift key by [⇒]. These two keys are used to attain the functions written above the keys. For example, [⇐] [ASIN] is the shifted [SIN] key and gives the arcsin function. We denote the four cursor keys (the keys with the triangles facing in the four directions) by [△], [◁], [▽] and [▷].

Your HP gives menu items in the bottom row of the screen. These are selected by pressing the corresponding letter key (A through F) below the desired menu item. We use the teletype font to indicate menu items, for example: [`EDIT`].

**Set Radian Mode**: We invariably want to measure angles in radians, not degrees. So turn on your calculator and see if the letters RAD show in the upper part of the display window. If not, press in succession the keys [⇐] [RAD] (shifted [MTH]). You can change back to degree mode with the same keystrokes (i.e., this is a "toggle" switch).

**Quitting**: Do you know how to turn your calculator *off*? **Answer**: Wait ten minutes without pressing any keys (wasteful of the battery)—or simply press [⇒] [OFF] (shifted [ON]).

**If You Get Confused**: While exploring your new calculator, you may get stuck in an unfamiliar mode. If this happens, press [ON]. When the calculator is already turned on, this key has the meaning "cancel" (this word is written beneath the ON key). You may have to do this a few times to return to a familiar situation.

## THE LAB

**General Instructions for All Lab Work**: Carefully record all your results and the parameters used to get them. Make sketches of the plots you make with the HP.

**Exercise 1** With the HP in radian mode, evaluate $\sin(\pi/3)$ *numerically*.
**Tip**: Use [⇐] [→NUM] (shifted [EVAL]) to convert the symbolic result to a numerical one. Check your answer by also computing the same quantity using an equivalent expression involving square roots that is derived from the famous 30-60-90 right triangle. Is there any discrepancy in the results?

**Exercise 2** Evaluate $\sqrt{2}$ numerically. How many "significant figures" does the HP give in "Standard" mode?

**Exercise 3** Define the functions $f(x) = x^2 + 1$ and $g(x) = x^3 - x^2 - 9x + 9$ and plot them on the same graph. From your graph, estimate the three values of $x$ where the plots cross.
**Tips**: To create the first function, key in ['] [α] [X] [$y^x$] [2] [+] [1] [ENTER]. Then type

[ ' ] [ $\alpha$ ] [ ⇐ ] [ F ] [ STO ] to register the function $f$ in memory (as lowercase f). Perform the analogous steps and register $g$ as g. To access the PLOT environment, press [ ⇒ ] [ PLOT ] (shifted [ 8 ]). Use the four cursor keys to navigate in the PLOT display. With the EQ (equation) field highlighted, use the [ CHOOS ] menu item to produce a sub-menu of registered functions. Use the cursor keys to highlight f and the [ CHK ] menu item to select it. Then the cursor keys and [ CHK ] to select g as a second function to be plotted on the same graph with f. Use the [ OK ] menu item to return to the main PLOT window. Highlight the H-VIEW field and key in the left horizontal plot limit as $-6$ (negative numbers are entered by using the change sign key: [ 6 ] [ +/- ]). Press either [ ENTER ] or [ OK ] to confirm this plot limit. The right horizontal field will now be highlighted—key in the value 6 and again press either [ ENTER ] or [ OK ] to confirm. Then highlight and [ CHK ] AUTOSCALE for the vertical plot limits. Use the [ ERASE ] and [ DRAW ] menu items to get the plot (it takes a few moments). When the graphs appear, use the [ (x,y) ] menu item and the cursor keys to locate the values where the curves cross. The cursor location is shown on the screen as a cross; at first it is hidden by the axes, so tap the cursor keys a few times till it comes into better view. Press [ CANCEL ] (i.e., [ ON ]) to return to the PLOT window ("cancel" is also the menu item [ CANCL ] in the picture window).

**Exercise 4** Form the "rational" function, $r(x) = g(x)/f(x)$ and plot it for $-20 \leq x \leq 20$. As usual, make a hand sketch of what you see. **Tip**: Register **g/f** as a new function **r** instead of entering $g$ and $f$ again.

**Exercise 5** Define the function $G(x) = x^3 + 2x^2 - 15x - 30$ and plot $G$ on several $x$-intervals to get a good idea of its behavior. **Tip**: The function $G$ registered as G and the earlier function $g$ registered as g show correctly in the PLOT environment [ CHOOS ] menu. However, a possible point of confusion is that all variables show as capital letters in the main [ VAR ] menu. Thus, in this menu, it will look as if there are two G functions.

a) Zoom in on each of the $x$-values for which $G = 0$. Does the function become "almost linear" as you zoom in? **Tip**: After using the [ (x,y) ] menu item and the cursor keys to locate the root, make a note of its approximate value, return to the Plot window ([ CANCL ] or [ CANCEL ]) and reset the horizontal limits to make a plot in a small interval about the root.

b) One root of $G(x) = 0$ is an integer, find that one. Verify that you have determined the correct root by factoring it out of $G$ (by hand). Then find the other two roots exactly and evaluate them numerically to check that your previous results were correct.

c) Define a new rational function, $R = G(x)/F(x)$, where $F(x) = x^4 + 3$. Decide if $R$ is "almost linear" for $x$ large (as was the case with the $r$ discussed earlier).

## AFTER THE LAB

**Exercise 6** Explain why the graphs of $g(x)$ and $r(x)$ are similar for "small" $x$.

**Exercise 7** The graph of $r(x)$ looks linear for "large" $x$. Can you figure out *what* linear function approximates $r$ for large $x$?

## BEFORE THE LAB

In high school, you learned the solution formula for the quadratic equation, $Ax^2 + Bx + C = 0$. However, you may not know any methods for solving higher degree or transcendental equations (i.e., equations involving trig functions, logs, etc.). Oftentimes, such equations can be approximately solved by "zooming in" on the desired root as you learned to do in the *Tutorial Exercises* Project and as you will do in the laboratory portion of this Project.

**Exercise 1** In each case, write an equation of the straight line $L$ that is described:

a) $L$ passes through $(1, 5)$ and is parallel to the line with equation $2x + y = 10$.

b) $L$ passes through $(-2, 4)$ and is perpendicular to the line with equation $x + 2y = 17$.

**Remark**: You may find it useful to check your text for the various forms used to describe straight lines: slope-intercept, point-slope, and two-point, to find the easiest one to use for the above problems.

**Exercise 2** You could make up a million problems like the ones in the previous exercise. It makes more scientific sense to solve *general* problems (though "warm-ups" with numerical values are often useful). Solve the following generalized version of the first part of Exercise 1:

> Write the equation of the straight line $L$ that passes through the point $(p, q)$ and is parallel to the line with equation $ax + by = c$. Check that your answer "works" with the specific numbers in part (a) of Exercise 1.

**Exercise 3** State the general problem corresponding to the second part of Exercise 1, solve it, and check with the specific example.

We will soon learn that:

| The slope of the tangent line to the parabola $y = x^2$ at $x = a$ is $2a$. |
|---|

Thus, the slope of this parabola at $x = 1$ is 2, the slope at $x = 3$ is 6 and so on. Please accept this as a fact for now.

**Illustrative Exercise:** Find the equation of the line through the point $P(2, 0)$ that is normal to the parabola $y = x^2$.

> **Solution**: Draw a figure (see Figure 1) that includes a rough sketch of the normal from the given point. Introduce the letter $a$ to denote the $x$-coordinate of the point where the normal cuts the parabola. Thus, the point $(a, a^2)$ is on both the parabola and the normal. The slope of the parabola at $x = a$ is $2a$, so the slope of the normal is the negative reciprocal value $-1/2a$. Equating this latter slope to the slope from $(a, a^2)$ to the given point $(2, 0)$ yields

$$-\frac{1}{2a} = \frac{a^2 - 0}{a - 2}$$

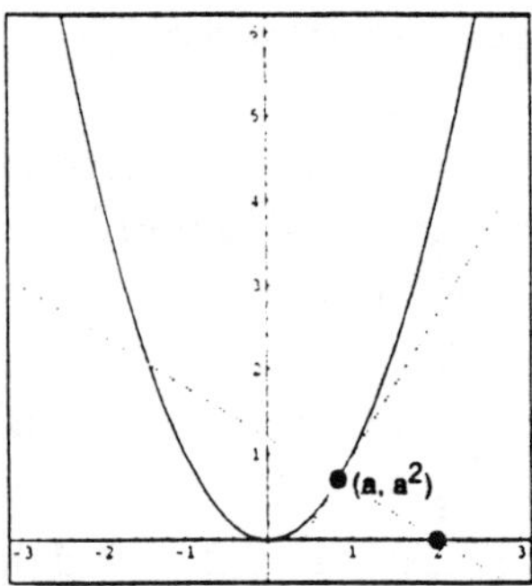


FIG. 1. Sketch for the Illustrative Exercise.

leading to the *cubic* equation,

$$2a^3 + a - 2 = 0$$

for $a$. We can see from our sketch that $a \approx 1$, but this is *not* the exact root (try it). Instead of resorting to trial and error, we can plot the function near $a = 1$ and then "zoom in" on the root. Successively plotting $2a^3+a-2$ on the intervals $[.7, 1.1]$, and $[.82, .85]$, we find that $a \approx .835$. Now is the time to be careful—we've gotten a good approximation to $a$, but that is *not* quite what the exercise called for! The correct solution to the problem is to produce the requested *normal* from $(2,0)$ to the parabola. This line is given by $\boxed{y = -\frac{1}{2a}(x-2), \quad \text{where } a \approx .835}$.

**Exercise 4** One problem solving maxim we've ignored in the above solution is that of trying to avoid the use of specific numbers. To maintain generality, let $(p, q)$ denote the given point (i.e replace $(2,0)$ by $(p, q)$). Show that the cubic equation giving the point $x = a$ where the normal from $(p, q)$ intersects the parabola $y = x^2$ is $\boxed{2a^3 - (2q-1)a - p = 0}$. It's good practice to check that this agrees with the specific case worked out in the Illustrative Exercise, so do it!

## THE LAB

**Exercise 5** In reference to the previous exercises, find the normal when $(p, q) = (4, 0)$. If you are uncertain how to proceed or how to present your solution, re-read the Illustrative Exercise.

**Exercise 6** In reference to the previous problems, when $(p, q) = (1, 17/4)$ show that there are *three* normals and determine them by doing three "zooms". Notice that the given point is *inside* the parabola.

**Notice how much more convenient and less error prone it is to use the *general* form for the cubic instead of setting up and simplifying the fractions for each specific point $(4, 0)$, $(1, 17/4)$, ....**

**Exercise 7** Show that the cubic equation in Exercise 6 has the *exact* solution $a = 2$. Factor your cubic to obtain a quadratic equation. Solve the quadratic to get all three roots of the cubic exactly. Check that the exact solutions agree with the ones you found by the zoom method.

## AFTER THE LAB

**Exercise 8** In light of your experience with the previous exercises. comment on the efficiency of the zoom method if highly accurate solutions (say 10 decimal place accuracy) are required.

**Exercise 9** Comment on the efficiency of the zoom method if solutions are required for *many* different points $(p, q)$.

**Exercise 10** In light of the results obtained in the lab exercises, a natural conjecture is that there are three normals for points interior to the parabola $y = x^2$ and only one for exterior points. Investigate the truth of this conjecture for the special case when the point is on the $y$-axis (that is, $p = 0$).

## BEFORE THE LAB

**Exercise 1** You land in your space ship on a spherical asteroid. Your partner walks 400 meters away along the smooth surface carrying a one meter rod, and thereby vanishes over the horizon. When she places one end on the ground and holds the rod straight up and down, you, lying on your stomach, can just barely see the tip of the rod. The ultimate quest is to determine the radius $r$ of the asteroid, but, for now, just draw a careful diagram and label it with the quantities you think will be useful in getting an equation involving only $r$ and known quantities.

**Exercise 2** After checking with your instructor that your diagram for the asteroid problem is correct, use it to derive the equation for the radius $r$ in terms of the given quantities. Again use symbols instead of specific numbers and check your equation with your instructor before going on with the project.

## THE LAB

### Making Tables with the HP

Sometimes it is best to study a function in a small region by making a table of its values there instead of a plot. The HP does not have a built-in command for making a table, so we will show you how to construct one. This construction requires "programming" the HP and we won't give full explanations—you can use a TV without understanding all the details of how a transistor works and the same is true for using your HP. It will help if you first read the short section "Entering and Executing Programs" starting on page 29-5 of the HP User's Guide. To enter a program begin with ⇐ ≪≫ (shifted −). This puts the HP in Program-entry mode. In this mode, you press SPC to separate consecutive variables or numbers and the right cursor key ▷ to move past closing delimeters like ' or ≫. **Remark**: If you already know how to enter programs and functions, you can skip over the following detailed keystroke oriented explanation and proceed to enter the programs directly from the listings given below.

Our first program is a utility that we call `L2V`. For those who are curious, we mention that the purpose of `L2V` is to convert lists to vectors—these vectors will ultimately form the columns of our tables. `L2V` uses the basic HP types, so rather than keying in them directly, we access them by the HP menus for the PRG key:

PRG TYPE OBJ→ →ARR.

Then press ENTER ' α. While holding the $\alpha$ key down, enter the name `L2V`. Release the $\alpha$ key and register the function by pressing STO. Return to the main menu by pressing VAR, where you should see a new menu entry for the `L2V` function.

**Remark**: We shall frequently have to enter strings of characters (such as the name `L2V` entered above). We will henceforth stop advising you to hold and release the $\alpha$ key and simply write the characters in place. See the User Guide, page 2-2, for a discussion of the various ways of entering characters.

We now construct the `TAB` program. `TAB` calls `L2V` as a "subroutine." Also, `TAB` uses the "local" variables A, B, and H, which respectively denote the first x-value to be tabled, the last x-value to be tabled. and the increment between x-values. At run time. these three variables are put on the stack before executing TAB (to be discussed further below). Begin entering the TAB function by keying in [⇐] [«»] [⇒] [→] (shifted [0]). To continue keying in TAB, type A [SPC] B [SPC] H to define the three local variables.

**Remark:** Henceforth we will usually omit showing the [SPC] explicitly (but you *need* to enter it on the HP to separate $\alpha$ strings!). In this style, the beginning of our code reads: [⇐] [«»] [⇒] [→] A B H.

Now enter the main function with the keystrokes:
[⇐] [«»] [ ' ] X [▷] [⇒] [→] X
[⇐] [«»] X X A B H SEQ [⇒] [→] XARG
which stores in the local variable XARG, the sequence of $x$-values where the function is to be evaluated. The TAB function continues:
[⇐] [«»] XARG L2V XARG F L2V 2
We now use the menus to access a special HP function:
[MTH] [MATR] [ROW] [ROW→].
Finally type
TRN
to conclude the TAB function (the four sets of closing braces are supplied automatically, but check that your editing hasn't created/lost any of them).

Register the function as TAB by first pressing [ENTER] [ ' ], entering the name TAB, and pressing the [STO] key. Return to the main menu by pressing [VAR], where you should see a new menu entry for the TAB function.

Now that you understand the special role of [SPC] and [▷] in Program-entry mode, we shall, in the future, write the code for such programs in the following style:

```
@ TAB--Create table with columns X and F(X)
@ Required globals:
@      F--DEFined function to be tabulated.
@ Local variables on stack:
@      A, B, H--first, last, step for x-values to be tabulated.
@ Other local variables:
@      X--independent variable of F.
@      XARG--list of the x values where the function is evaluated.

TAB:
<< -> A B H                            @ read these local variables from stack
   << 'X' -> X                         @ set up independent variable
      << X X A B H SEQ -> XARG         @ store x values in local variable XARG
         << XARG L2V                   @ convert x values to row vector
            XARG F L2V                 @ apply F to x values; convert to row
            2 ROW->                    @ convert the two rows to a matrix
            TRN                        @ transpose to get 2 column array
         >>                            @ end scope of XARG
      >>                               @ end scope of X
   >>                                  @ end scope of A, B, H
>>                                     @ end procedure
```

Notice that we use the symbol @ to denote a comment (neither the @ nor the following comment are actually typed when entering a program). Also note that we have written the code in an indented format for readability on the printed page. There is no need to indent it the same way when entering it into the HP. In fact, such indentation will *not* be permanently stored in any case. On the other hand, it *is* sometimes convenient to start a new line on the HP display when entering the code. To do this, use the right shifted [·] key.

As the program notes indicate, we must have a function stored as F before the TAB function is run. Here, we choose the function $f(x) = \cos x - x$. To enter and register this function, use the keystrokes.
['] F [⇐] [( )] X [▷] [⇐] [=] [COS] X [▷] [−] X [▷] [⇐] [DEF].
Notice that in registering a function F(X), we use DEF; while in registering an expression F (e.g., for plotting) we use STO.

**Remark:** In the future, we will merely write 'F(X)=COS(X)-X' to denote the process of defining a function.

**Tip:** Functions applied to lists use the [+] operation in a special way (see section 17-2 in the User Guide). Since your function might later be applied to a list, you should avoid using [+] in your functions (unless you intend this special usage). You have three ways to do this: rewrite the function so you don't need [+], use ADD in place of [+], or use [-] [-] in place of [+].

At last we are ready to use TAB: put the numbers 0, 1.5, .2 on the stack and press the menu item [TAB] (i.e., 0 [ENTER] 1.5 [ENTER] .2 [TAB]). To conveniently explore the resulting table, enter the EDIT environment ([⇐] [EDIT]) and use the cursor keys to scroll up and down. Press [CANCEL] when done. The initial portion of the table should look something like this:

```
[[ 0 1]
 [ .2 .780066577841 ]
 [ .4 .521060994003 ]
 [ .6 .22533561491 ]
```

Scrolling down reveals the remainder of the table:

```
 [ .8 -.103293290653 ...
 [ 1  -.459697694132 ]
 [ 1.2 -.837642245523...
 [ 1.4 -1.2300328571 ...
```

**Exercise 3** Give an intuitive explanation of why there is a solution of $\cos x = x$ in the interval $0.6 \leq x \leq 0.8$ and use the TAB command to determine this solution with error less than 0.05.

**The Asteroid Problem—continued**

**Exercise 4** You now know the correct equation for $r$. Using plotting and/or tabling, find the solution to one decimal place accuracy for the given numerical values. **Remark:** Your equation probably contains an addition (+) operation, so take note of the **Tip** above!

**Comment:** How much would you contribute to the Indigent Math Professors' Support Fund for an easily derived estimate of the solution to within 1 meter? Actually, this fee is covered in your tuition for this course—stay tuned. Without questioning that plotting and tabling are terrific tools, your experience with the current problem may cause you to wonder if there aren't some better methods of solving equations. If so, you will be glad to hear that you will, indeed, be learning about more advanced solution methods this semester.

## AFTER THE LAB

We will be returning to the asteroid problem several times in later projects, so it is worthwhile to reflect on your experience so far:

**Exercise 5** In light of your investigations, discuss how you could have gotten the answer more efficiently.

**Exercise 6** In light of your investigations, discuss the virtues and limitations of plotting and tabling in solving equations.

## BEFORE THE LAB

Your text shows that the derivative of $f(x) = x^2$ is $f'(x) = 2x$. This is a special case of the **Power Law:**

$$f(x) = x^n \Longrightarrow f'(x) = nx^{n-1} \qquad \text{for } n = 1, 2, \ldots .$$

### The START loop

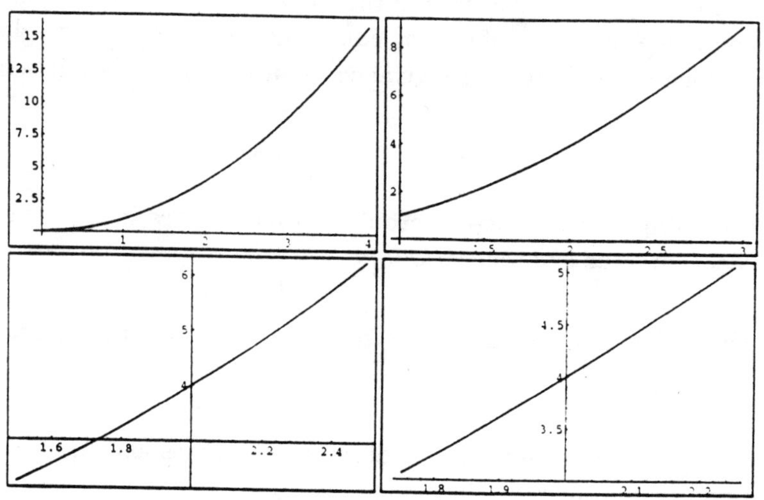


FIG. 2. First four zooms of $x^2$ at $x = 2$.

It is tiring to generate a series of "zoom" PLOTs one at a time when you know in advance which point you are interested in. To relieve the tedium, we give you the LPLOT procedure for automatically zooming in on $f(x)$ near the point $x = a$. LPLOT requires that you store an algebraic expression for X in the EQ variable. As an example, we prepare to zoom in on $f(x) = x^2$ by entering: [ ' ] X [ $y^x$ ] 2 [ENTER] [ ' ] EQ [STO]

**Remark**: From now on, we assume that you know how to handle the [ $\alpha$ ] key and so we just write the letters without saying "hold the $\alpha$ key and type EQ," etc.

Here is the LPLOT procedure:

```
@ LPLOT--Show series of zooms on function near X = A
@ Required globals:
@       EQ--'algebraic' expression to be plotted.
@ Stored variables:
@       X PPAR from PLOT.
@ Local variables:
@       A--limiting point.
@       H--the tabulation step.
@       N--the number of zooms.

LPLOT:
<< -> A H N                 @ read these local variables from stack
  << 1 N START              @ start of zoom loop (5 zooms)
        H 2 / 'H' STO       @ set interval to half previous interval
        FUNCTION            @ access plot environment
        'X' INDEP           @ set variable to X
        A H - A H + XRNG    @ set plot x range to [A-H,A+H]
        AUTO                @ set plot y range to AUTO
        ERASE DRAW          @ clean window and draw next zoom
     NEXT                   @ end of zoom loop
     PICTURE                @ redraw final zoom and show menu
  >>
>>
```

To zoom in on $f(x) = x^2$ near $x = 2$, put 2 (the `A` value), 4 (the `H` value) and 5 (`N`, the number of zooms) on the stack. Then press the **LPLOT** menu item. On each pass through the **START** loop, a plot is generated with the plot range varying from $x = a-h$ to $x = a+h$. The quantity $h$ is set to 4 before the loop. As we enter the loop, it is cut in half, so the first plot goes from $x = a - 2$ to $x = a + 2$. Before the second plot, $h$ is cut in half again, so that the second plot goes from $x = a - 1$ to $x = a + 1$. Similarly, the third plot goes from $x = a - 1/2$ to $x = a + 1/2$ and so on. Since $a$ is set to 2 before the loop, the actual successive numerical plot ranges are $[0, 4]$, $[1, 3]$, $[1.5, 2.5]$, .... So we are, indeed, "zooming" in on $x = 2$. Your first four zooms should look like Figure 2 (of course you don't get the axes labeling in the HP window) and the fifth zoom is shown in Figure 3.

**Tip**: Input a small number (like 2) for `N` until you are *sure* the loop is doing what you want.

**Exercise 1** Use the output of the final zoom shown in Figure 3 to verify that the slope of $f(x) = x^2$ at $x = 2$ is what it should be. The plot does not have equal scales on the axes, so you should use the *numbers* on the axes for your computations, *not* the apparent slope. Reproduce the figure as a rough sketch and draw any necessary lines on your sketch. Make the best estimate you can of the slope and also state the error in your approximation to the exact slope. **Warning**: Take into account the fact that the origin in the figure is not where the axes cross—that is, be sure to compute the difference quotient, not just $y/x$.

This project will also require you to plot several functions on the same graph, so review the procedure for this as given in the *Tutorial Exercises* project.

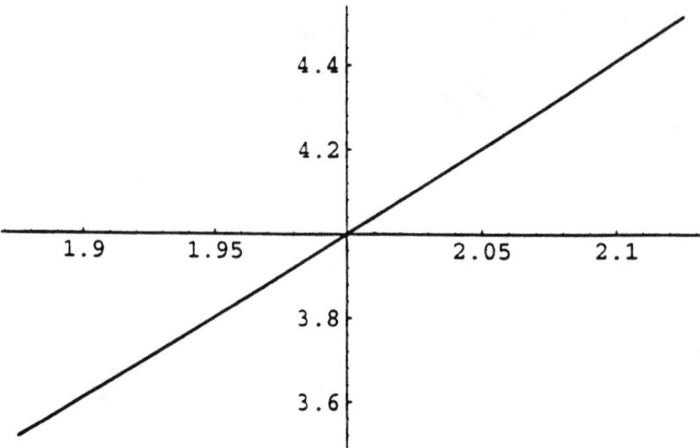


FIG. 3. Fifth zoom of $x^2$ at $x = 2$.

## THE LAB

**Exercise 2** Use the HP to check the power law for $n = 3$ and $n = 4$ at $x = 2$. Method: Set up a defined function of the three variables $x$, $h$, and $n$ for the difference quotients and evaluate them at $x = 2$ and small $h$ values, say $h = 0.001$ and $h = -0.001$.

**Exercise 3** In this exercise, we find the tangent line to a curve using the idea that the tangent line is the *limit* of secant lines. As a nontrivial example, we use the function:

$$f(x) = \frac{(x^3 - 5)(x^2 - 1)}{x^2 + 1}$$

and find the tangent line at the specific point $(2, f(2)) = (2, 1.8)$.

To this end, consider two points on the graph, namely $(2, f(2))$ and $(b, f(b))$, where $b$ is close to, but not equal to 2. The slope of the "secant" line (the line that connects the two points) is $(f(b) - f(2))/(b - 2)$. Let's start with $b = 4$. To calculate the secant line, set up the DEFined function `F(X)=(X^3-5)(X^2-1)/(X^2+1)`. Then `STOre` 4 into `B`, the slope `(F(B)-F(2))/(B-2.0)` as `SLOPE`, the secant `F(2.0)+SLOPE*(X-2.0)` into `SECANT` (only **`SECA`** will show in the menu label) and finally store `F(X)` as **`FUNC`**.

Simultaneously display the graphs of the stored **`FUNC`** and its secant line **`SECANT`** near the point of interest. Proceed, taking $b$ closer and closer to 2.0 until you have what looks like a tangent to the curve at $(2, f(2))$—that is, find a value of $b$ (close to, but not equal to 2.0) such that you effectively have a tangent. Determine $f(b)$ for each $b$ that you use, and sketch the plot of the curve and the "secant". At what value of $b$ close to 2.0 does the secant begin to look like a true tangent?

**Exercise 4** Use the **`LPLOT`** procedure as in Exercise 1 to approximately determine the slope and tangent line to each of $f(x) = x^3$ and $f(x) = x^4$ at the point $x = 2$. Use the cursor keys and the `(x,y)` menu item to get two points near $x = 2$ for computing the approximate slopes.

**Exercise 5** Use the method of the previous exercise to again solve Exercise 3, that is, find the tangent line to $f(x) = (x^3 - 5)(x^2 - 1)/(x^2 + 1)$ at $x = 2$ by zooming in on $x = 2$.

**Exercise 6** Use the method of the previous two exercises to explore the slope and tangent line to

$$f(x) = (x-2)^{2/3} + 2x^3$$

at $x = 2$. You will need more zooms than previously to unearth the behavior near $x = 2$—in particular, what does the slope appear to be after 4, 8 and 12 zooms? Continue zooming until you are sure you've captured the true microscopic behavior near $x = 2$. Is there a unique tangent line at $x = 2$? Unique slope at $x = 2$? Why do you think so many zooms were needed?

**Tip:** Fractional exponents are tricky—if you don't do it right, you can wind up with complex (i.e., non-real) answers. Here is one way that works: `'XROOT(3,X-2)^2+2*X^3'`

**Exercise 7** Find the (approximate) slopes of $f(x) = \sin x$ at the points $x = \pi/3$ and $x = \pi/6$.

## AFTER THE LAB

**Exercise 8** Use the power rule (as cited above) to check that the numerical values you found for the slopes of $x^3$ and $x^4$ were approximately correct.

**Exercise 9** We haven't yet derived the formula for the derivative of $\sin x$, but look it up in your textbook and check the results you got in Exercise 7.

## BEFORE THE LAB

**Exercise 1** Consider the following piecewise defined function:

$$f(x) = \begin{cases} x^3+2 & \text{if } x \le -1; \\ x^4+1 & \text{if } x \ge 1; \\ x^2+x+1 & \text{elsewhere.} \end{cases}$$

a) Find out where $f(x)$ is continuous and discontinuous and justify your conclusions.

b) Determine values $b$ and $c$ such that the related function

$$f(x) = \begin{cases} x^3+2 & \text{if } x \le -1; \\ x^4+1 & \text{if } x \ge 1; \\ x^2+bx+c & \text{elsewhere.} \end{cases}$$

is continuous everywhere.

## THE LAB

We begin with two problems that review and extend some HP material.

**Exercise 2** DEFine a function `'F(X) = X^3 + X^2'`. To use such a defined function of a single variable, just type the argument and press the F menu key. Here, for example, [1] [F] should produce the answer 2. Report the results that you get for:

```
F(2)
F(T)
F(CAT)
```

Also use the **TAB** function given in the *The Asteroid Problem* project to table `F(X)` for `X = 1, 2, 3, 4, 5`. **Hint:** To avoid trouble with TAB, recall the **Tip** about not using + in defined functions applied to lists.

**Remark:** When you DEF a function, the result is stored as a procedure whose local variables are the arguments of the function—DEF is just a convenient shorthand to allow use of a convenient function notation. The above function winds up being stored as the procedure

```
<< -> X    'X^3+X^2'
>>
```

Notice the "algebraic" notation. Another way to code this procedure using "stack" notation would be

```
<< -> X
   << X 3 ^          @ Here ^ stands for the y^x key, so we get X^3
      X 2 ^          @ Similarly, get X^2
      +              @ Add the two powers
   >>
>>
```

The stack notation follows what we would type from the keyboard. but the algebraic notation is easier to code and read. Indeed. when we use stack notation. we will always accompany such code with explanatory comments.

**Exercise 3** To what extent can you replicate the results of the previous exercise with a STOred variable G given by `'X^3 + X^2'`? Recall that this is the way we code "functions" for the PLOT environment. **Hint**: Use EVAL.

**Exercise 4** Find

$$\lim_{x\to 0} \frac{\sqrt{25+3x}-\sqrt{25-2x}}{x}$$

by each of the following methods:

a) Make a plot of the fraction near the point $x = 0$. This often "works" even when, as in this case, the function is not defined at the critical point.

b) Make TAB tables on either side of $x = 0$ (again, be sure to change that + sign to −−).

c) The straightforward use of TAB doesn't "home in" on the critical point. Below is an alternate LTAB code that is more tuned for this purpose.

```
@ LTAB--Create table with columns X and F(X) near X = A (to the left)
@ Required globals:
@       F--DEFined function to be tabulated.
@ Local variables on stack:
@       A--limiting point.
@       N--the number of table entries.
@ Other local variables:
@       X independent variable of F.
@       XARG is a list of the x values where the function is evaluated.
@ Subroutine:
@       L2V converts a list to the corresponding vector.
@
@ To get RTAB just delete the change sign (NEG) command.

LTAB:
<< -> A N                          @ read these local variables from stack
  << 'X' -> X                      @ set up independent variable
    << X X 1 N 1 SEQ               @ get {1 2 ... N}
       1 <<2 SWAP ^>> DOSUBS       @ get {2 4 ... 2^N}
       INV NEG A ADD -> XARG       @ invert, negate and add A to each element
       << XARG L2V                 @ convert x values to row vector
          XARG F L2V               @ apply F to x values; convert to row
          2 ROW->                  @ convert the two rows to a matrix
          TRN                      @ transpose to get 2 column array
       >>                          @ end scope of XARG
    >>                             @ end scope of X
  >>                               @ end scope of A, N
>>                                 @ end procedure
L2V:
<< OBJ-> ->ARRY >>                 @ Put list on stack and convert to vector
```

**Remark:** `INV` results from using the $\boxed{1/x}$ key and `NEG` from the $\boxed{+/-}$ key. As always, you can alternately type the actual letters. **Tip:** Use the `MEMORY` environment to make a copy of `TAB` called `LTAB`. Then use the `EDIT` environment to make the few needed changes.

Try this code out by entering the limiting value and number of zooms:
[0] [ENTER] [6] [LTAB].

As it stands, the code explores the *left* limit; make a modified copy called `RTAB` for exploring the right limit by simply removing the change sign command in `LTAB`. Report the results of running your modified code.

**Exercise 5** Evaluate the following limits as $x \to 0$ using any one of the methods introduced so far:

a) $\dfrac{x^3 - x^2 - 4x + 4}{x - 1}$

b) $\dfrac{\sin x}{x}$ (This is an important limit.)

c) $\dfrac{1 - \cos x}{x}$ (So is this.)

d) $\dfrac{\sin 5x}{x}$

e) $(1 + x)^{1/x}$ (Another important limit.)

**Exercise 6** Verify the assertions you made in Exercise 1 by doing a simultaneous plot of `'X^3 + 2'` and `'X^2 + X + 1'` on the interval $[-2, 0]$ and using the [`(x,y)`] menu item to check the continuity at $x = -1$ by seeing if the curves cross at $x = -1$. Hand in a sketch of the plot you see, labeling the point $x = -1$. Similarly, check the continuity at $x = 1$ by plotting `'X^4 + 1'` and `'X^2 + X + 1'` on $[0, 2]$. Repeat for the second part of Exercise 1.

## AFTER THE LAB

**Exercise 7** Algebraic limits of the 0/0 form can be simplified to a form where the limit can be obtained by just evaluating the function. Sometimes, standard algebraic tools can be used advantageously. Use the trick of multiplying top and bottom by a sum of square roots ("rationalizing the numerator") to algebraically check the result of Exercise 4 and check that your Lab results were correct.

**Exercise 8** If, as claimed above, "algebraic limits of the 0/0 form can be simplified to a form where the limit can be obtained by just evaluating the function", what about

$$\lim_{x \to 8} \frac{x^{2/3} - 4}{x - 8}?$$

**Hint:** Factor (i.e., divide) the quantity $x^{1/3} - 2$ from both numerator and denominator.

**Exercise 9** You have noticed that many limits can be obtained by just evaluating the given function at the given point. State precisely a condition under which this is valid. Which of the following limits can be evaluated in this simple way? Justify your answers.

a) $\lim_{x \to 5} \dfrac{x+5}{x^4+x^2+1}$

b) $\lim_{x \to 5} \dfrac{x^2-25}{x-5}$

c) $\lim_{x \to 5} \sqrt{x^6-5}$

**Exercise 10** The function in Exercise 1 seems artificial. Can you think of any cases in science where a function would naturally be defined piecewise?

## COMING ATTRACTIONS

It would be misleading to leave this project without pointing out that the methods presented above have limitations. Consider the limit

$$\lim_{x \to 0} \frac{\sin(\tan x) - \tan(\sin x)}{x^7}.$$

Entering 0 and 10, **RTAB** gives:

```
0.5           -0.0447304...
0.25          -0.0358329...
0.125         -0.0339382...
0.0625        -0.0347390...
0.03125       -0.0309237...
0.015625      -0.4398046...
0.0078125     -0
0.00390625    720.5759...
0.00195312    -92233...
0.000976562   -11805...
```

A close look should arouse your suspicions. We seem to be approaching something like $-.03\ldots$ and then suddenly we get some erratic behavior, a drop to 0, and then utter nonsense. What's going on? Well, calculators (and even computers) can't do *exact* arithmetic and thus suffer from round-off error. You've just seen a particularly egregious form of this ailment called "catastrophic subtractive cancellation." Later on in the calculus, we'll cover the topic of Taylor series. Taylor approximations provide a satisfying resolution of the problem with this (and many other) limits. The small $x$ behavior of our function is provided by the following Taylor approximations:

$$\sin(\tan x) \approx x + \frac{x^3}{6} - \frac{x^5}{40} - \frac{55\,x^7}{1008} - \frac{143\,x^9}{3456},$$

$$\tan(\sin x) \approx x + \frac{x^3}{6} - \frac{x^5}{40} - \frac{107\,x^7}{5040} - \frac{73\,x^9}{24192}.$$

Subtracting and dividing by $x^7$ shows why the *exact* limiting value is $-1/30 \approx -0.033333$.

## BEFORE THE LAB

The Binomial Theorem is:

$$(a+b)^n = a^n + \frac{n}{1}a^{n-1}b + \frac{n\cdot(n-1)}{1\cdot 2}a^{n-2}b^2 + \ldots + b^n. \tag{1}$$

**Exercise 1** There are "dots" in the above expression for the binomial theorem. To be sure that you understand the correct pattern, write the first missing term explicitly (i.e. the one with the $b^3$ in it). Also write the last missing term explicitly (i.e. the one with the $a^1 = a$ in it).

**Exercise 2** Show that your answers to the previous exercise are correct by expanding out $(a+b)^n$ for $n = 1, 2, 3$ and 4.

**Exercise 3** Write the binomial theorem for the special case when $a = 1$ and $b = x$. This is the form we'll use below.

**Exercise 4** By factoring out $a^n$, show that (with a slightly different definition of $x$), the general binomial theorem in Equation 1 can be written in the form of the previous exercise (so it isn't so special after all).

**Exercise 5** Some of the founders of calculus were intrigued by the possibility of using the binomial theorem for *fractional* exponents. Considering only "small" $x$, try to discover what value of the parameter $a$ is appropriate in

$$\sqrt{1+x} \approx 1 + ax?$$

**Hint**: Square both sides. Does your value of $a$ agree with the binomial theorem for $n = 1/2$? Assuming that the binomial theorem for integer $n$ is a reliable guide for fractional $n$ too, write the next term of the approximation.

**Exercise 6** For integer $n$, the binomial theorem has a finite number of terms (exactly $n+1$ terms, in fact). Assuming that the theorem makes sense for fractional $n$, do you think that there is a hard and fast last term? Explain.

### More on Loops

In the *Derivatives, Slopes and Tangent Lines* project, we used a simple START loop that just repeats a portion of code a certain number of times. More typically, a loop uses a "counter" (here K). Here is a "stack" code implementation (SQ results from [⇐] [$x^2$]):

```
<< -> A B H          @ start, end and step in loop
   << 0              @ initialize sum to zero
      A B            @ copy loop limits
      FOR K          @ start K loop
         K SQ +      @ square element and add to sum (stack notation)
      H STEP         @ increment counter and go to top of loop
   >>
>>
```

Suppose we store this procedure under the name SUMSQ. Then entering 1, 7, and 2 and pressing the menu item SUMSQ computes $1^2 + 3^2 + 5^2 + 7^2$ (that is, K starts at 1 and is incremented by 2 repeatedly until it gets up to 7).

If you prefer algebraic notation (^ is $y^x$):

```
<< -> A B H                @ start, end and step in loop
   << 0 -> S               @ initialize sum
      << A B               @ set loop limits
         FOR K             @ start K loop
            'S+K^2' EVAL   @ square and update sum (algebraic notation)
            'S' STO        @ store sum
         H STEP            @ increment counter and go to top of loop
         S                 @ put final sum on the stack
      >>
   >>
>>
```

## THE LAB

**Exercise 7** From your "Before the Lab" work, you know that using only two terms of the binomial theorem gives $(1+x)^5 \approx 1+5x$. Is this really a usable approximation? One way to start exploring this question is to PLOT the two functions $f(x) = (1+x)^5$ and $g(x) = 1+5x$ on the same graph for the three intervals $0 \leq x \leq 0.1$, $0 \leq x \leq 1.0$, and $0 \leq x \leq 10.0$. What about *negative* $x$?

**Exercise 8** Investigate the accuracy of $\sqrt[3]{1+x} \approx 1 + \frac{1}{3}x$ for small $x$ along the lines of Exercise 7.

**Exercise 9** Investigate the "infinite series",

$$S = 1 + \frac{1}{2^1} + \frac{1}{2^2} + \frac{1}{2^3} + \dots .$$

Try to approximate $S$. You might use code like this (INV is $1/x$):

```
<< -> N                @ end of loop
   << 1                @ initialize sum to one
      1 N
      FOR K
         2 K ^ INV +
      NEXT             @ step is 1, so use NEXT instead of STEP
   >>
>>
```

Start your investigation with the loop limit N as 10.

**Exercise 10** Similarly investigate the series,

$$S = 1 + \frac{1}{2^2} + \frac{1}{3^2} + \frac{1}{4^2} + \dots .$$

**Caution:** For this problem, the previous code needs to be changed in *two* places.

**Exercise 11** Similarly investigate the series.

$$S = 1 + \frac{1}{2} + \frac{1}{3} + \frac{1}{4} + \dots .$$

## AFTER THE LAB

**Exercise 12** Consider the infinite series,

$$S = 1 - 1 + 1 - 1 + 1 - 1 + \dots .$$

If we group like this:

$$S = (1 - 1) + (1 - 1) + (1 - 1) + \dots,$$

then we get $S = 0$. On the other hand, if we group it like this:

$$S = 1 + (-1 + 1) + (-1 + 1) + (-1 + 1) + \dots,$$

we get $S = 1$. We seem to have created something (1) out of nothing (0), what do you think about this?

**Exercise 13** It seems reasonable to ask if the binomial theorem works for *negative* exponents. Investigate the case $n = -1$. Have you seen this somewhere before?

**Exercise 14** Evaluate the limit

$$\lim_{x \to 0} \frac{\sqrt{25 + 3x} - \sqrt{25 - 2x}}{x}$$

from the *Limit & Continuity* project by using the idea described in Exercises 4 and 5.

## BEFORE THE LAB

Taking derivatives with the HP is similar to plotting:

1) Invoke the symbolic environment with [⇒] [SYMBOLIC].
2) Move the cursor down one position to the "Differentiate..." item.
3) Press the [OK] menu entry.
4) Use the [EDIT] menu entry to create a quoted expression in the EXPR field, say, 'X^3'.
5) Move the cursor down to the VAR field and enter X (or whatever dependent variable you are using).
6) The RESULT field probably already reads "Symbolic" (use [CHOOS] if it doesn't).
7) Press the [OK] menu entry.
8) The derivative will appear as a quoted expression on the stack ('3*X^2' for the function in step 4) that you can STOre in a VARiable if you will need it later .

Alternatively, you can define an expression beforehand by STOring an expression in a VARiable, say, 'X^3' ENTER 'F' STO' and use the [CHOOS] menu item to select F in step four above. Or—even faster—store the function in the EXPR variable and skip step four altogether. **Tips:** You have to set the VAR field every time you select the Symbolic "Differentiate..." item. The actual derivative of

$$\frac{1}{1+\sqrt{1+\sqrt{x}}}$$

is

$$\frac{-1}{4\left(1+\sqrt{1+\sqrt{x}}\right)^2\sqrt{1+\sqrt{x}}\,\sqrt{x}}.$$

The HP expression for this derivative can be STOred and EVALuated, but, in all honesty, it is hard to read. The best you can do is to press the [VIEW] key and see that the HP expression is a double compound fraction equivalent to the answer just cited. Press [CANCEL] (twice) to exit VIEW-mode. Another example: if you use the HP to differentiate

$$(1+(1+x)^2)^3, \tag{2}$$

then you will see:

```
1:  '2*(1+X)*3*(1+(1+X)
    ^2)^2'
```

on level one of the stack. To simplify this expression, STOre it back into the 'EXPR' variable and again invoke the SYMBOLIC environment. Move the cursor *up* one position and select the "Manip expr..." item (i.e. manipulate expression). Since we've already stored our expression in EXPR, we can immediately use the [COLCT] menu item and press [ENTER] to get the nicer form of the result given by:

```
1:  '6*(1+(1+X)^2)^2*(1
    +X)'
```

Again, get a better look by pressing [VIEW] to see the simplified form:

$$6 \cdot \left(1+(1+X)^2\right)^2 \cdot (1+X)$$

**Remark:** Here's how to construct the tangent approximation to a function: STOre the original expression (say the function in Equation (2)). as, say, F and the resulting derivative as, say, FF. STOre the $x$ at the point where the tangent is desired in the variable X (say the value 2). Then [F] [EVAL] gives $f(2)$ (1000 in our example) and [FF] [EVAL] gives $f'(2)$ (1800 in our example). Then the tangent is $y = f(2) + f'(2)(x-2) = 1000 + 1800(x-2)$.

**Exercise 1** Verify each of the above differentiation results by hand.

## THE LAB

**Exercise 2** Use the HP to differentiate each of: $\sin^2 x$, $\sin x^2$, and $\sin(\sin x)$. Note that each of the three functions is *different*—be careful to distinguish the first two: $\sin^2 x = (\sin x)^2$, but $\sin x^2 = \sin(x^2)$. Assume for now, without proof, that $D \sin x = \cos x$ (this is correct for $x$ measured in radians). Use the chain rule and this important differentiation result to check the HP output.

**Exercise 3** Use the HP to compute the tangent line to the function

$$f(x) = x^4 - 4x^3 + 3x^2 - 2x + 1$$

at $x = 2$. Check the result by hand.

**Exercise 4** Use the HP to compute the derivative of $(2x^2-1)(x^3+2)$. Write down the simplest form you can get. Use the product rule to check that your output makes sense.

**Exercise 5** Use the HP to compute the derivative of

$$\frac{x}{x-2}.$$

Write down the simplest form you can get from the HP.

**Exercise 6** Compute the derivative of $(x+x^2)^5(1+x^3)^2$. Write down the simplest form you can get from the HP. **Tip:** You can use the cursor keys to scroll while in VIEW-mode.

**Exercise 7** Use the HP to compute the derivative of

$$\frac{1}{1-2/x}.$$

Write down the simplest form you can get from the HP.

**Exercise 8** Pick three nasty looking differentiation problems for which you have answers in the back of your text. Do plain differentiation with the HP and then try to simplify the result into a form close enough to the text answer so that the answers can easily be compared. Be sure to state both answers if you don't completely succeed in making them agree. (Be aware that the "back of the book" answer is occasionally wrong!)

## AFTER THE LAB

**Exercise 9** The fully simplified results in Exercises 5 and 7 should be identical—why?

## BEFORE THE LAB

In previous projects, we've solved equations numerically by plotting and/or tabling using a zoom method. We've encountered two problems: (1) getting an initial estimate of the root location and (2) getting high accuracy once the root *is* located approximately.

Indeed, plotting and/or tabling themselves can often help with the first problem. Also, if the problem comes with a physical context, we can sometimes make an intelligent guess at the root location. Later in the course, we'll see that approximating the given equation with a simpler one (e.g., by series expansions) can also be of help in getting an approximate root location.

In this project, and the next several, we study the second problem: refining the value of the root, given a crude approximation. Our principal tool will be Newton's Method, but here we begin with the Bisection Method. As we will see, the Bisection and related methods are complimentary to the methods related to the Newton Method (called Newton-Raphson in some books).

Why don't we just settle for the previous methodology of "zooming" with plotting and/or tabling? Well these are often competitive for a single fixed equation, but they suffer from the defect of requiring intensive supervision by the user. This defect rules out the plotting methodology when many equations have to be solved or when high accuracy is required.

**Prerequisite**: Read about the method of bisection in your text.

**Exercise 1** What is the connection between the *Intermediate Value Property* and the bisection method? (Clear explanation, please.)

**Exercise 2** The folklore says that it takes about 10 bisection steps to gain 3 decimal places of accuracy. For example, if we know the root to within an integer, then 10 bisections will determine 3 places after the decimal point. The folklore is right—explain why.

As a "warm up" problem, we consider finding the positive solution of the equation $x^2 = 5$. Since square roots can be found by pressing a button on a calculator, you may well ask, "why bother with an analytic method for finding square roots?". There are several answers:

- It *is* just a warm up problem.
- It's good practice to have a "benchmark" problem as a check on new methodology (e.g., for checking that your computer code is correct).
- This problem has aesthetic appeal since it dates back at least 2000 years to the great Babylonian mathematicians.

**Exercise 3** As a consequence of the result in Exercise 2 it takes about 4 bisections to gain *one* decimal place. Prove that the square root equation, $x^2 = 5$, has a root in the interval $[2, 3]$, and then do 4 steps of bisection by hand to get an improved approximation to the desired square root. You get *no* credit for a series of cryptic equations—you must explain your successive calculations and results. Neatness counts! Use of clear English sentences is required!

## THE LAB

It is surprisingly difficult to write a correct bisection method code (unfortunately it is all too easy to write one that we *think* is correct, but isn't). Here is a basic HP version:

```
@ BISECT--Recursive implementation of Method of Bisection
@ Required globals:
@      F--DEFined function to be solved (F(X)=0).
@ Stored variables:
@      m--latest root estimate.
@ Local variables:
@      A--left endpoint.
@      B--right endpoint.
@      EPS--the accuracy requirement.
@ Limitations:
@      The program assumes the root is bracketed, i.e., F(A)*F(B)<0
@      The program assumes that A<B

BISECT:
<< -> A B EPS                        @ read these local variables from stack
   << '(A+B)/2' EVAL 'm' STO         @ bisect interval & store midpoint
      IF 'B-A<EPS' THEN              @ check accuracy
         m                           @ if OK put m on stack & exit
      ELSE                           @ else
         IF 'F(m)*F(A)>0' THEN       @ which half interval has root?
            m B EPS BISECT           @ set up stack and call BISECT on [m,B]
         ELSE                        @ else
            A m EPS BISECT           @ set up stack and call BISECT on [A,m]
         END                         @ end the half interval test
      END                            @ end the accuracy test
   >>
>>
```

Here, the equation being solved is $f(x) = 0$, $a$ and $b$ are values of $x$ such that $f(a)f(b) < 0$, and *eps* is the desired accuracy. Thus, to get one decimal place accuracy for $\sqrt{5}$, assuming that you had DEFined `'F(X)=X^2-5'` and STOred the above program under the name BISECT (only the letters BISEC show in the VAR menu) you would key in:

```
2 3 .05
```

and press the BISECT menu entry. Try it! You should get 2.234375 whose square is 4.99243....

In this Lab, we will want to know the number of bisections required to reach the requested accuracy. Thus, edit your code to include this facility as follows:

```
@ BISECT--Recursive implementation of Method of Bisection (Tutorial Version)
@ Returns:
@      number of iterations in level 2 (N)
@      approximate root in level 1 (m)
@ Required globals:
@      F--DEFined function to be solved (F(X)=0).
```

```
@ Stored variables:
@      m--latest root estimate.
@      N--number of bisections (Store a a zero in N before running).
@ Local variables:
@      A--left endpoint.
@      B--right endpoint.
@      EPS--the accuracy requirement.
@ Limitations:
@      The program assumes that 0 has been stored in N
@      The program assumes the root is bracketed, i.e., F(A)*F(B)<0
@      The program assumes that A<B

BISECT:
<< -> A B EPS                       @ read these local variables from stack
   << '(A+B)/2' EVAL 'm' STO        @ bisect interval & store midpoint
      'N+1' EVAL 'N' STO            @ update iteration counter
      IF 'B-A<EPS' THEN             @ check accuracy
         N m                        @ if OK put N and m on stack & exit
      ELSE                          @ else
         IF 'F(m)*F(A)>0' THEN      @ which half interval has root?
            m B EPS BISECT          @ set up stack and call BISECT on [m,B]
         ELSE                       @ else
            A m EPS BISECT          @ set up stack and call BISECT on [A,m]
         END                        @ end the half interval test
      END                           @ end the accuracy test
   >>
>>
```

**Tip:** This code requires that you STOre 0 in the global variable `N` before running it. The first time you do the usual ritual: [0] [ENTER] ['] N [STO]. You could do that before *every* run, but once the variable `N` exists, here's an easy way to set it back to 0: [0] [⇐] [N]. Here, the last keystroke means to press the `N` menu item.

**Remark:** If you are interested in programming, you could try to put in a test to make *sure* that `A<B` and `F(A)*F(B)<0`. One idea would be to return some unusual number like 99999 if the test fails. A better one would be to consult the *Advanced User's Reference Manual* for the methodology of printing user error messages.

**Exercise 4** Run the second version of the BISECT code. Compare the number of bisection iterations it *really* takes to get one decimal accuracy with the assertion of Exercise 3. Finally, use this code to get the result correct to 3 decimal places.

**Exercise 5** Recall that an approximate solution to the "asteroid equation,"

$$\cos\frac{s}{r} = \frac{r}{r+h},$$

with $h = 1$, and $s = 400$ places $r$ in the interval $[80000, 80001]$. Use the method of bisection to obtain 3 decimal accuracy.

**Exercise 6** Repeat the last exercise to obtain 6 decimal accuracy.

**Exercise 7** *Prove* that with the parameters $s = 500$, $h = 1$, the asteroid equation has a root in the interval $[125000, 125001]$ and then use bisection to get 3 decimal accuracy.

## AFTER THE LAB

**Exercise 8** If we want to find $\sqrt{5}$, why do we use the equation $x^2 = 5$ instead of $x = \sqrt{5}$?

**Exercise 9** Compare the actual number of bisection iterations it takes to achieve the target accuracy with each of the Lab Exercises with the assertion of Exercise 2.

## BEFORE THE LAB

**Prerequisites:** Please read (or re-read) the "Before the Lab" section of the *Bisection Method* project (skipping over the exercises). In particular, note the justification for studying the simple problem of determining $\sqrt{5}$, even though you can get a good approximation by a simple key press of your HP. Also, read the material in your text about Newton's method for solving the equation, $f(x) = 0$, paying special attention to the derivation of the Newton iteration,

$$x_{n+1} = x_n - \frac{f(x_n)}{f'(x_n)},$$

for replacing the "current" approximation $x_n$ by the (hopefully) improved approximation $x_{n+1}$.

**Exercise 1** For the specific equation, $x^2 = c$, where $c$ is a given constant, use Newton's method to derive the Babylonian square root iteration scheme of averaging the current guess (say $a$) and $c/a$ to obtain

$$\frac{1}{2}\left(a + \frac{c}{a}\right)$$

as the new approximation. Using the starting approximation $x = 2$, carry out two iteration steps by hand for the case $c = 5$. Explain your calculations clearly and check the accuracy of your result.

**Remark:** If $a > \sqrt{c}$, then $c/a < \sqrt{c}$ and *vice versa*. This is the motivation behind the Babylonian method.

We now discuss the implemention of Newton's method in the HP in the context of determining a numerical approximation for $\sqrt{5}$. Thus, consider the function $f(x) = x^2 - 5$. Begin by STOring an expression for this function in the variable F. A plot (see Figure 4) reveals that $\sqrt{5} \approx 2.2$: You can check this with your HP by making a plot of F and using

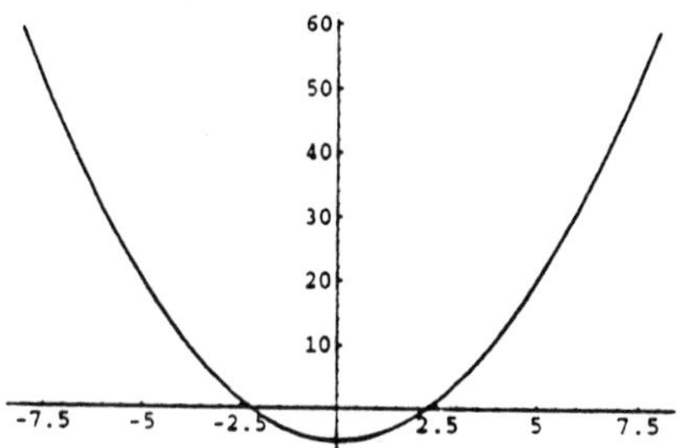


FIG. 4. Result of plotting F on the interval $[-8, 8]$.

the x,y menu option and the cursor keys.

Let's illustrate geometrically how Newton's method generates the first iterate. To get an easily intepretable figure, we use the (poor) starting value of $x = 6$. Figure 5 shows that the tangent at $(6, f(6))$ cuts the $x$-axis between 6 and the true root, thus producing an improved $x$-estimate of $\sqrt{5}$ of about $x = 3.5$.

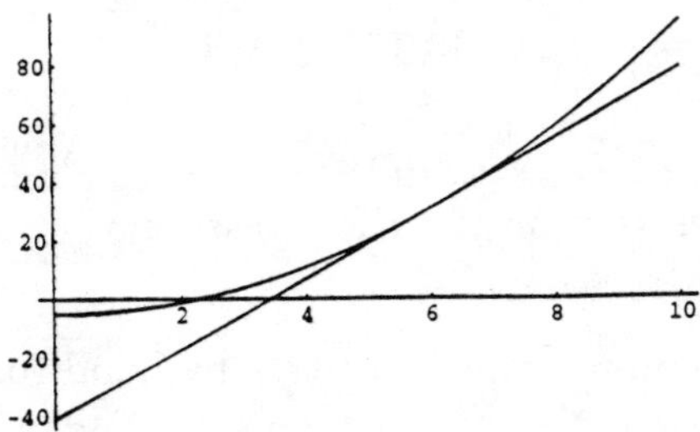


FIG. 5. Geometric demonstration of the first Newton iteration, the poor starting value $x = 6$ is improved to around $x = 3.5$ by this tangent line approximation.

## Computing derivatives from the keyboard with the HP

In the *Derivatives with the HP* project, you learned how to compute derivatives using the `SYMBOLIC` environment. It is also possible to compute them directly from the keyboard. The methodology is to [ENTER] successively the function to be differentiated, and the independent variable and then press the [⇒][∂] key (shifted [SIN]). For example, assuming that you do *not* have a variable `X` in the current HP directory, you can get the derivative of our function $x^2 - 5$ with:

['] X [$y^x$] 2 [−] 5 [ENTER]
['] X [ENTER]
[⇒][∂]

This produces `'2*X'` at level 1.

**Remark:** If you *do* have a variable `X` with a numerical value, then you'll get the derivative evaluated at this point.

**Tips:** Since you already `STO`red this `F` above, you can shortcut the process like this:

[⇒][F]
['] X [ENTER]
[⇒][∂]

To delete a stored variable (say `X`) from the keyboard:

['] X [ENTER]
[⇐][PURG]

You can also use the [MEMORY] environment to purge variables interactively.

A procedure to differentiate a `STO`red expression `F` with respect to `X` would be simply:

```
<< 'X' PURGE F 'X' ∂ >>.
```

If you wanted to store the derivative in the variable `DF` instead of showing it at level one of the stack, just add a `STO` command:

```
<< 'X' PURGE F 'X' ∂ 'DF' STO >>.
```

In these two short routines, the `'X' PURGE` fragment guarantees that there is no stored variable `X` and hence that the derivative will be a symbolic expression and not just a number.

## THE LAB

To assess the accuracy of the $\sqrt{5}$ approximations you'll develop shortly, we first use the built in square root operation on the HP. Pressing 5 $\boxed{\sqrt{x}}$ yields 2.2360679775 and this serves as a standard for comparisions.

If you were willing to compute the derivative by hand (or use the above routines) and DEF functions $f(x)$ as F and $f'(x)$ as DF, Newton's method on the HP could be as simple as STOring:

```
<< -> X 'X-F(X)/D(X)' ->NUM >>
```

and this method, or a variation of it, will be useful to you in later projects where it is better (or necessary) to compute $f'$ yourself. But in this project, we'll use an HP implementation of Newton's method which computes the derivative internally. Note that in the following listing, we used the lower case word 'deriv' to indicate the $\boxed{\Rightarrow}$ $\boxed{\partial}$ keystrokes that evoke the derivative operation (this was done because of font limitations).

```
@ NEWT--Interactive implementation of Newton's Method
@ Required globals:
@      F--stored expression for the function to be solved (F=0).
@ Stored variables:
@      DF--numerical value of the derivative of F.
@ Local variables:
@      X--initial guess at root.
@ Limitations:
@      Newton's method only converges locally in general

NEWT:
<< -> X                  @ read this local variable from stack
      F 'X' deriv        @ numerical derivative; "deriv" is right-shift-SIN key
      'DF' STO           @ store numerical derivative in DF
      'X-F/DF' ->NUM     @ compute x - f(x)/f'(x) in algebraic notation
   >>
>>
```

**Exercise 2** $\boxed{\text{ENTER}}$ the starting value 6 and press the NEWT menu key corresponding to the HP procedure given above. You should obtain a value consistent with Figure 5. To get the *next* iterate, just press NEWT again. And so on. Try it for a total of 4 or 5 iterations with the starting value 6.0. Report on the accuracy attained.

**Exercise 3** Now try an experiment with taking an even poorer starting value, $x = 15.0$. How many iterations does it takes to converge to $x = 2.23607$?

**Exercise 4** To automate repeated iterations of Newton's method we use a START loop; in the following version, we "hardwire" the number of iterations at eight.

## Newton's Method

```
@ NEWT--Loop implementation of Newton's Method showing iterates
@ Required globals:
@      F--stored expression for the function to be solved (F=0).
@ Stored variables:
@      DF--numerical value of the derivative of F.
@ Local variables:
@      X--initial guess at root.
@ Limitations:
@      Newton's method only converges locally in general

NEWT:
<< -> X                  @ read this local variable from stack
   << 1 8 START          @ do 8 iterations
         F 'X' deriv     @ numerical derivative; "deriv" is right-shift-SIN key
         'DF' STO        @ store numerical derivative in DF
         'X-F/DF' ->NUM  @ compute x - f(x)/f'(x) in algebraic notation
         'X' STO         @ store answer as new guess
         X               @ put iterate on stack
      NEXT               @ do next iteration
   >>
>>
```

To show the approach to convergence, this loop puts each of the eight successive iterates on the stack. You will see the last 4 iterates on the HP screen. To see the earlier iterates, use the [△] key. Try out the loop with the initial value 16 and then alter it to answer the question: how many iterations does it take to get 10 correct significant figures with the (absolutely insane) starting value 200?

**Exercise 5** Use the above program or a variant to compute the square roots of the seven equally spaced numbers, 0.5, 1.0, 1.5, 2.0, 2.5, 3.5, and 4.0 to 6 significant figures. Use the starting value 1.0 all seven times.

**Exercise 6** This Project concludes with the study of the more challenging function

$$f(x) = (x^3 - 2.1x^2 + x - 2)/(x^6 + 1)$$

a) *Prove* that $f$ has a zero in the interval $[-10, 10]$

b) Graphically show that $f$ has only one zero in this interval. **Hint**: after doing an ordinary PLOT on $[-10, 10]$, repeat setting the vertical plot limits to $[-.1, .1]$ to closely examine the region near the $x$-axis.

c) Pick a reasonable starting $x$-value, e.g. the closest integer, and do enough Newton iterations to get at least six significant figure precision. Copy down and hand in the HP output showing all the iterates.

d) Now pick $x$ bigger; e.g. $x = 3.0$ or larger. Do several iterations and explain the results.

e) Try $x$ around 0.5 and tell what happens, and why it happens.

f) Try $x$ around -1.0 and tell what happens, and why.

## AFTER THE LAB

**Exercise 7** In light of your experience with Newton's Method, what caution(s) need to be observed?

**Exercise 8** Suppose we want to write a general and efficient routine to find the square root of a given positive number $a$. As you know, computers deal naturally with binary expansions. One approach is to use this fact to efficiently compute $n$ and $b$ such that $a = 4^n b$ with $0 < b < 4$. Then $\sqrt{a} = 2^n\sqrt{b}$, so we only need a square root routine for numbers between 0 and 4. Since we don't intend to be around when people use our routine, we need to to use a *fixed* starting value $x_0$ and a *fixed* number of iterations $niter$. Suppose we select the fixed value $x_0 = 1.0$ as we did in Exercise 5. Pick and justify, on the basis of your Lab experience, a likely value for $niter$. How would you make sure that your $niter$ was the best possible choice?

**Caveat:** It is ***not*** being asserted that the algorithm outlined above is actually the way square roots are computed on any particular hardware. Defining such basic algorithms is a dicy business that depends on the given hardware and on an analysis of how sophisticated a starting guess to use, whether to store precomputed tables, and so forth. So, you see, computing square roots isn't so trivial after all—if ***you*** are the person designing the algorithm that determines what happens when the calculator button is pressed. If this intrigues you, take a numerical analysis course sometime.

## BEFORE THE LAB

### Applications of Newton's Method

**Law of Reflection.**—The law of reflection states that the angle of incidence $\theta_1$ is equal to the angle of reflection $\theta_2$, that is $\theta_1 = \theta_2$ or, equivalently, $\alpha = \beta$ (see Figure 6). Furthermore, if the speed in the medium is constant, then the reflection path from a point $P$ to a point $Q$ consists of two straight line segments meeting at the reflection point as in the Figure.

**Exercise 1** Given two points $P$ and $Q$, we want to determine the location of the point of reflection on the interface (say $x$ measured horizontally from one of the points). As in Figure 6, assume that the two points are separated from each other by a total horizontal distance $L$ and that their vertical distances from the reflecting interface are respectively $a$ and $b$. Use the law of reflection to derive an expression for $x$ in terms of $a$, $b$ and $L$. **Hint:** Find a *simple* equation by using similar triangles. **Check:** For $a = 50$, $b = 25$, and $L = 150$, the correct answer for $x$ is 100.

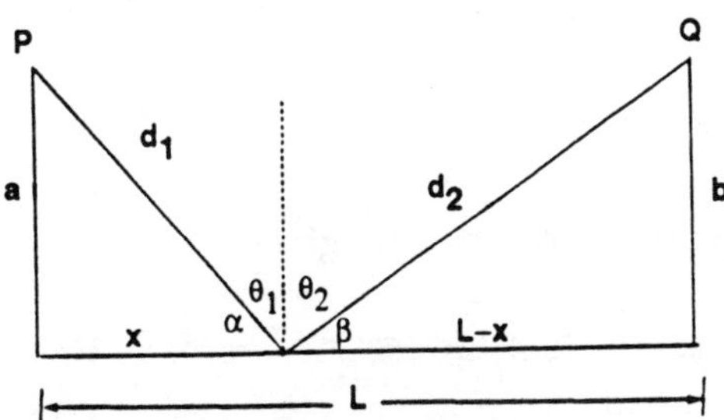


FIG. 6. Geometry of reflection.

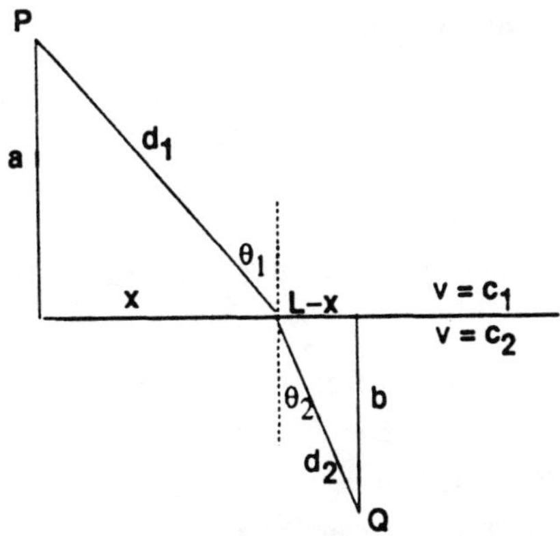


FIG. 7. Geometry of refraction.

**Snell's Law.**—The law of refraction (often called Snell's law) states that for an interface separating two media, the angle of incidence $\theta_1$ is related to the angle of refraction $\theta_2$, by the relation

$$\frac{\sin\theta_1}{c_1} = \frac{\sin\theta_2}{c_1}$$

(see Figure 7). Here $c_1$ and $c_2$ are the respective speeds above and below the interface and $\theta_1$ and $\theta_2$ are again the angles measured from the normal. Once more if the speeds are assumed constant, then the refraction path consists of two straight line segments meeting at the refraction point.

In contrast to the reflection case where the reflection point can be found explicitly, the location of the refracting point $x$ involves a fourth degree polynomial:

$$(c_2^2 - c_1^2)x^4 - 2L(c_2^2 - c_1^2)x^3 + (L^2(c_2^2 - c_1^2) + c_2^2b^2 - c_1^2a^2)x^2 + 2Lc_1^2a^2x - L^2c_1^2a^2 = 0 \quad (3)$$

In real applications, this "ray tracing" problem becomes even more difficult (and often takes super-computing capability) because many layers are involved. Moreover, in many applications, it is necessary to deal with complications such as allowing the speed to vary in each layer or considering distinct wave types simultaneously (e.g., elastic "P-waves" and "S-waves"), etc. Ray tracing plays an important part in many fields ranging from geophysics to creating cinematic special effects.

By the way, Equation 3 can be derived as follows: Observe that Snell's Law, Figure 7 implies

$$\frac{1}{c_1}\frac{x}{d_1} = \frac{1}{c_2}\frac{L-x}{d_2}.$$

Cross multiplying, squaring and transposing gives

$$(c_2\, x\, d_2)^2 = (c_1(L-x)d_1)^2.$$

Then the intrepid may square out and simplify to get the result.

**Asteroid Problem.**—The previous work on this problem showed that the radius $r$ of the asteroid is governed by the equation

$$\cos\frac{s}{r} = \frac{r}{r+h},$$

where $s$ is the distance the traveling astronaut walks and $h$ is the length of the rod that she holds vertically for the stationary ground-level astronaut to sight on.

**Exercise 2** Recall that for the particular values $s = 400\,\text{m}$ and $h = 1\,\text{m}$, you've shown that $r \approx 80,000\,\text{m}$. To avoid "wasting" the 5 significant figures involved in writing "80,000", make the substitution $r = R + x$ in the asteroid equation (here $R = 80000$, but for goodness sakes don't write out "80000" three times, write "$R$").

## THE LAB

**Snell's Law continued.—**

**Exercise 3** Show that for the values $a = b = 1$, $L = 4$, $c_1 = 1$, and $c_2 = 1/2$, Equation 3 for the location of the refraction point reduces to

$$3x^4 - 24x^3 + 51x^2 - 32x + 64 = 0.$$

This latter equation actually has *two* real roots:

a) Hand in a graph that shows their approximate locations.

b) Explain why only one of the roots is relevant to the refraction problem.

c) Find the relevant zero to six significant figures. Show the result of each Newton iteration.

d) Provide a graph showing an interval including the relevant zero and the tangent line from your starting $(x, f(x))$ intersecting the $x$-axis at the first iterate.

**Exercise 4** Solve for the relevant zero if the parameters are all the same as in the last problem except that now $c_2 = 2$. Add up the answers to this problem and the previous one–can you explain this sum?

**Asteroid Problem continued.—**

**Exercise 5** So far, we've applied Newton's method to polynomials, but it applies just as well to transcendental equations such as the one you obtained in Exercise 2. Obtain the relevant root $x$ (not $r$!) to 4 significant figures with Newton's method. State the corresponding solution for the physical quantity $r$.

**Exercise 6** When you previously found the "physically sensible" solution to the asteroid problem you may have noticed that there were a lot (infinitely many, actually) of roots $r$ in the interval $[0, 400]$. Find the largest two roots ($r$-values) that are less than $s = 400$ to 6 significant figures and explain what would have happened to the astronauts if the asteroid really had a size corresponding to the radii given by these two smaller roots. Similarly, explain what happens for an asteroid with radius given by the yet smaller roots.

## AFTER THE LAB

**Exercise 7** Derive the law of reflection from Fermat's principle of least time: light will follow the path that minimizes the time. You may assume that the paths are straight lines. **Hints:** Since the speed is constant, this is equivalent to minimizing $d_1 + d_2$ in Figure 6. After taking a derivative, interpret the result in terms of the angles $\alpha$ and $\beta$. You could also determine $x$ from this equation, but hopefully you found a simpler way in Exercise 1.

## BEFORE THE LAB

Nothing to do beforehand this time. We know that Newton's method usually converges quickly, if we start "sufficiently close" to a simple root. In this Project, we examine some problems with the method. Don't get paranoid, we are going out of our way to create difficulties—in most cases, Newton's method is a wonderful tool.

## THE LAB

**Exercise 1** Run the usual Newton iteration loop on the function

$$f(x) = 27x^6 - 216x^5 + 387x^4 - 440x^3 + 408x^2 - 224x + 48$$

with starting value 1.0.

a) How many Newton iterations does it take to get 5 significant figure accuracy?

b) Plot $f$ for $-2 < x < 2$ and try to explain why the Newton method is unusually slow.

**Chaos Theory.**—In the remaining problems, we apply Newton's method to the simple function $f(x) = x^3 - x = x(x-1)(x+1)$ with the roots $x = -1$, 0 and 1. If the initial value $x_0$ is chosen close to any one of these three roots, the Newton iterations converge quickly to that root. However, if one selects $x_0$ on a certain interval (it isn't near any root), strange things happen. This behavior is related to an area of mathematics called "chaos" theory. Chaos theory has received great attention during the past 10 years and there was even a best seller about it[1]. The term "chaos" arises from the fact that for some problems (e.g., weather prediction), *small* changes in the starting conditions make *large* changes in the long term behavior. Thus, long term prediction becomes impossible, despite knowing the exact equations.

**Exercise 2** Two important $x$-values in our study are $s3 = 1/\sqrt{3}$ and $s5 = 1/\sqrt{5}$. One would *not* want to pick $x_0$ as either $\pm s3$. Why not? Also, what happens if one picks $x_0 = s5$? That is, what are $x_1, x_2 \ldots$?

**Exercise 3** Make PLOTs of $f$ in both the critical region $[-1, 1]$ and on a larger scale, say $[-4, 4]$.

a) Explain from the plots why you would *expect* that an initial value $x_0 > s3$ would lead to $x_n \to 1.0$ as $n \to \infty$. (And by symmetry, $x_0 < -s3$ leads to $x_n \to -1.0$ as $n \to \infty$).

b) Similarly, explain why if $|x_0| < s5$, you would expect $x_n \to 0$.

c) Numerically verify the assertions in parts a and b.

**Exercise 4** (The "chaos" part). Here are provided a list of starting values $x_0$ lying between $s5$ and $s3$ and getting closer and closer to $s5$. Using each starting value, run 15 or so Newton iterations and carefully describe what you observe (including any sign changes in the Newton iterates).

---

[1] *Chaos* by James Gleick, Viking-Penguin Inc., 1987.

a) use $x_0 = 0.448955$

b) use $x_0 = 0.447503$

c) use $x_0 = 0.447262$

d) use $x_0 = 0.447222$

e) use $x_0 = 0.447215$

f) use $x_0 = 0.447213$ (whoops!)

## AFTER THE LAB

**Exercise 5** Discuss the results obtained in Exercise 4 with respect to "sensitivity" to the choice of initial value.

**Exercise 6** Suppose that it is known that for a certain function $f$, a unique root exists on the interval $[a, b]$. We've seen that in some unusual cases. Newton's method can either fail or can converge too slowly. On the other hand, settling for a "safe" method like bisection *all the time* is too expensive. How could we automatically monitor when Newton's method is in "trouble"? Casually discuss "rescuing" Newton's method in such cases.

## BEFORE THE LAB

Now that we understand how Newton's Method works, we look at several equation solvers that are built in to the HP—of course, Newton's method is a key element in their construction!

We discuss three powerful methodologies built into the HP. Polynomial equations are markedly simpler than the general case and the HP has a polynomial root finder devoted to this special case that delivers all the roots. There is also a general equation finder that requires additional information—a starting guess (because it uses Newton's method as a key component). Lastly, we'll discuss the HP's powerful interactive graphical solution technique.

### Finding all roots of polynomial equatic s

As an example, study the cubic equation $x^3 - 3x^2 + 1 = 0$. The most efficient way to solve is to use the HP's polynomial root-finder. Enter the HP's interactive equation solver by pressing [⇒] [SOLVE] (shifted 7) and then highlight `Solve poly...` and press [OK]. In the `COEFFICIENTS` field, enter brackets ([⇐] [[]]) and then the coefficients of the polynomial separated by spaces (use the [SPC] key). The highest order terms come first and you must supply zeroes for any "missing" powers. In our example, your input should look like:

```
[1 -3 0 1]
```

Then press [ENTER] and the menu item [SOLVE] and the roots will appear in the the `ROOTS` field. Note that *complex* roots are possible (try `[1 0 4]`) and these are presented in the notation `(a, b)` standing for $a + bi$. In our example cubic equation, the roots are all real and you should see:

```
[ -.532088886238 .6...
```

To see the rest of the roots use the [EDIT] item. Also, a labeled ("tagged") copy of the roots is put on the stack. Thus, you can examine the roots by exiting the `SOLVE` environment (with [CANCEL]) and either using [EDIT] or [⇐] [VIEW]. But a nicer way to get a readable list of the roots on the stack is to [CANCEL] out of `SOLVE` and enter the command `DTAG` to strip off the tag. After using `DTAG`, you can separate the roots by pressing [PRG] and the menu items [TYPE] [OBJ→]. You get the roots at levels 2, 3, ... and the number of roots at level 1. Still another approach is to use [OBJ→] once to separate the tag from the roots and then [DROP] the tag and use [OBJ→] again to separate the roots on the stack.

### Finding a real root of a general equation or expression

For non-polynomial equations such as the "asteroid" equation, enter the `SOLVE` environment and select the first entry, `Solve equation`. In the `EQ` field, enter the equation, here, `'COS(S/R)=R/(R+H)'`. You will then see entries for each of the symbolic quantities, `S, R, H`. In the `S` field enter the known arclength 400, in the `R` field enter the starting value 80000, and in the `H` field, enter the known height of the stick 1. Then highlight the unknown field R and press the menu item SOLVE. You will soon see the more accurate answer 80000.8306827 in the R field. On CANCELing the SOLVE environment, you will see a tagged version of the answer on the stack at level 1. Once again, you can use the DTAG command to strip off the tag if you wish. If you prefer, you can enter an expression instead of an equation and the HP assumes that you want a zero on the right side, for example SOLVEing 'COS(S/R)-R/(R+H)' is equivalent to SOLVEing 'COS(S/R)-R/(R+H)=0'.

Remark: Finding a root to a general equation is nontrivial and you should not blithely accept calculator or computer output without being sure that a root actually exists near the presented value. For example, because of limited precision, your calculator or computer may not be able to detect the difference between a function crossing the $x$-axis or just coming very close to it. Depending on your degree of cautiousness, the SOLVE environment provides several tools for assessing whether a root has actually been found.

1) After pressing SOLVE and getting a "solution," press the INFO menu item. A window pops up showing the root value again and saying either:

   **Zero** The equation is satisfied exactly to within the HP's 12 significant digit precision.

   **Sign Reversal** The equation changes sign very close to the estimated "root," but there is no HP floating number where the equation is satisfied to 12 significant digits. This is not necessarily bad news—it often just means that the root "fell in the crack" between successive floating point numbers that exist on the HP. But it *might* mean that at points between these successive floats one side or the other of the equation doesn't exist (e.g., becomes infinite or complex).

   **Extremum** The function comes close to the $x$-axis, but may not cross it. For example, for a small positive value of $\epsilon$, $x^2 + \epsilon$ exhibits this behavior: there is an extremum (minimum) at $x = 0$, but not a root there.

2) If you highlight the EQ field and use the EXPR= menu item, then the left and right brackets of the "root" are displayed on the stack.

3) You can watch the signs of the intermediate approximations by pressing ENTER immediately after pressing SOLVE.

**Remark:** You should be aware that the value of the root that the HP finds will vary slightly according to the initial guess (try 79000 instead of 80000). Using a computer, one finds that even though the HP delivers 80000.8306827 as the "answer," a still better answer is 80000.833. That is, not all the digits in the HP "answer" are correct. In this regard, here are some computer-based (from *Mathematica*) results worthy of study:

```
s = 400; h = 1;
f[r_] := Cos[s/r] - r/(r+h)

f[80000.8306827]    (* HP answer *)
                 -13
  -4.12781 10
```

```
f[80000.833]         (* best 8 sf answer *)
                   -14
   -5.07372 10

f[80000.8333070797] (* computer answer---Mathematica *)
                   -15
   -2.66454 10
```

Note that while we cannot trust all the digits printed by the HP, the answer returned *does* give a very small value to the function—these two *different* meanings of the word "solution" should be kept in mind, i.e., equation satisfied to so many significant digits vs. root accurate to so many significant digits.

### Graphical equation solving

First re-examine our example cubic equation, $x^3 - 3x^2 + 1 = 0$. Invoke the PLOT environment and enter 'X^3-3*X^2+1' in the EQ field. Let us suppose that a preliminary plotting investigation has revealed that the roots are all in the interval $[-1, 3]$ and set the plot limits accordingly. We have already learned how to use the (x,y) menu item with the cursor keys to get approximate roots. Here, this methodology reveals roots near .54, .66, and 2.9. Please verify this before going on. Then press NXT to get the menu back and use the menu item FCN. Then move the cursor to one of the roots and press the menu item ROOT. You will see a high accuracy value for the root. Move to the other two roots and again NXT ROOT. On CANCELing out of the PLOT environment (i.e. two cancels), you will see the three roots on the stack. Of course, in general, we would not be able to get all the roots pictured in a single plot and would, instead, employ several plots to find all the roots of interest.

Next treat the asteroid equation by entering 'COS(400/R)=R/(R+1)' and plotting say in [79500, 80500]. When the HP is given an *equation* to plot, it plots the expressions that appear on both sides of the equal sign. When the plot is finished, move the cursor near to the point of intersection and FCN ISECT. These menu options produce the *intersection* of the two curves and once again we get $r = 80000.83\ldots$ which is seen on the stack when you exit PLOT.

**Exercise 1** A 100-ft tree stands 20 feet from a 10-ft fence. The tree is then "broken" at a height of $x$ feet, so that it just grazes the top of the fence and touches the ground on the other side of the fence with its tip (see Figure 8). Derive the equation $x^3 - 68x^2 + 1100x - 5000 = 0$ for $x$.

## THE LAB

**Exercise 2** Use the HP's polynomial root-finder to get numerical values for the quadratic equation, $x^2 - 3x + 4 = 0$.

**Exercise 3** Find the five roots of the polynomial equation $x^5 - 3x + 4 = 0$.

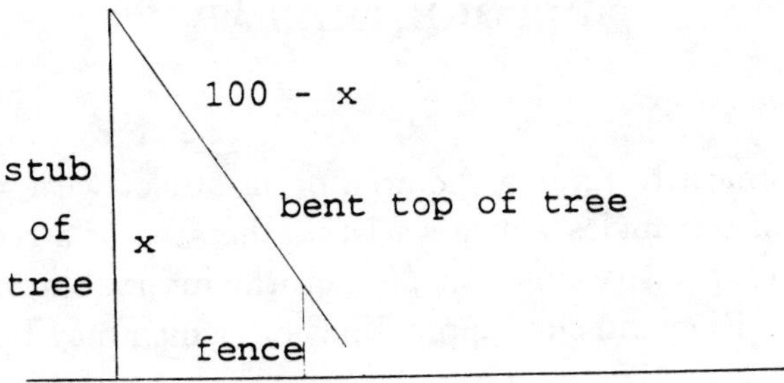


FIG. 8. Diagram for Exercise 1.

**Exercise 4** Solve the "broken tree" equation and decide which roots make physical sense.

**Exercise 5** Use the HP to solve the cubic equation. $x^3 - 2ax^2 - 4x + 8a = 0$ for several values of the parameter $a$. On the basis of the results you get, make a conjecture about the roots for any value of $a$ and prove that you are right.

**Exercise 6** Plot $\tan x + x$ on $[0, 12]$ to get crude values for its positive zeroes near the origin. Use the HP's PLOT environment to get accurate values for the first 4 positive roots of the equation $\tan x + x = 0$. Check the accuracy of your solutions.

## AFTER THE LAB

Nothing this time.

## BEFORE THE LAB

### Introduction

A fundamental approach to finding the area of an object with general shape is to approximate it by sums of rectangles (since we *know* the area of a rectangle). For the area between the $x$-axis and a positive function $f(x)$ on the interval $[a, b]$, your text introduces the idea of a systematic Riemann sum approximation, which may be written as

$$\text{Area} \approx R = \sum_{i=0}^{n-1} f(x_i^*)\,\Delta x_i, \tag{4}$$

where $x_i^*$ is a point in the interval $[x_i, x_{i+1}]$ and $\Delta x_i$ is the length of this interval (i.e., $x_{i+1} - x_i$). In this project we will use this idea to approximate some interesting areas. We choose the specific Riemann sum obtained using *equal* subdivisions $\Delta x_i = \Delta x = (b-a)/n$ and the *midpoints* $x_i^* = a + (i + 1/2)\Delta x$, so that the sum approximating the area is given by

$$\text{Area} \approx R = \sum_{i=0}^{n-1} f\left(a + (i + 1/2)\Delta x\right)\,\Delta x, \qquad \Delta x = (b-a)/n.$$

Figure 9 shows an approximation of this type for a small value of $n$.

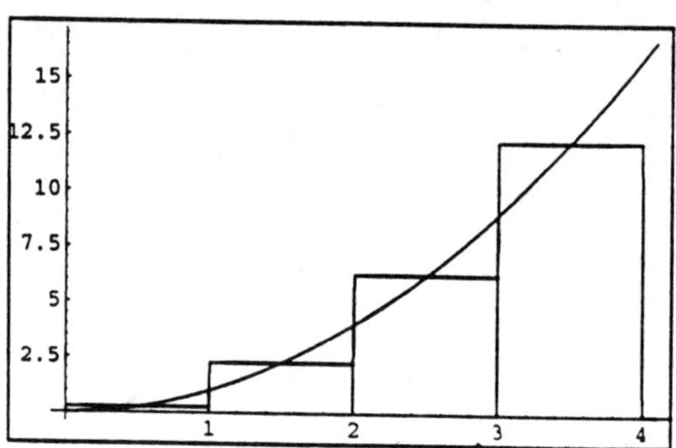


FIG. 9. Function and rectangular area approximation.

**Exercise 1** Equation 4 is often written as

$$R = \sum_{i=1}^{n} f(x_i^*)\,\Delta x_i,$$

where now $x_i^*$ is a point in the interval $[x_{i-1}, x_i]$ and $\Delta x_i$ is the length of this interval (i.e., $x_i - x_{i-1}$). Explain why this equation expresses the same quantity.

### Area of a circle

You know that the area of a circle of radius $r$ is $A = \pi r^2$, where $\pi \approx 3.14$. But until now, you have probably just accepted this dictum. It is easy to argue from similarity that the area is *proportional* to $r^2$. And it is easy to give the constant of proportionality a *name* like $\pi$. What is *not* easy is to give a method for accurately computing $\pi$. But before pursuing that, let's start with the crude approach of simply drawing a circle on a grid and counting the squares enclosed (sometimes we have to get results in the absence of a good theoretical framework).

**Exercise 2** Count the squares totally enclosed by the circle in Figure 10 and add in an estimate of the portion of squares on the boundary that are also inside to get an estimate of the area of the circle. Use your results to estimate $\pi$.

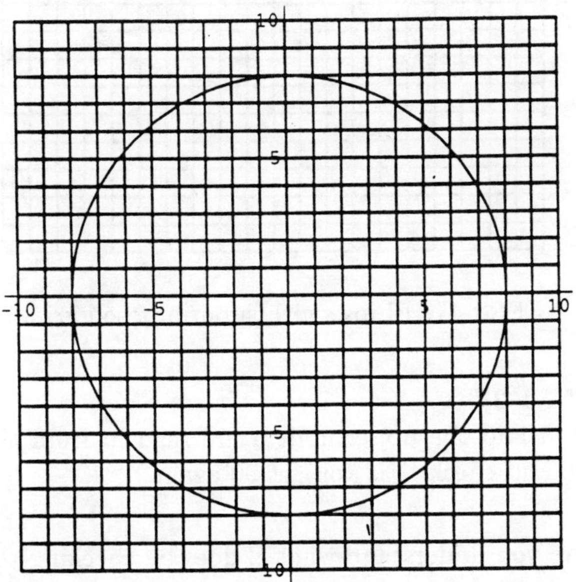


FIG. 10. Circle with superimposed grid.

**Exercise 3** Use Figure 11 to obtain an estimate of the area of the ellipse shown there.

**Exercise 4** We have implicitly *defined* the constant $\pi$ as the area of the unit circle. Use *only* this definition to show that $\pi$ is exactly four times the area under the curve $y = \sqrt{1-x^2}$ on the interval $[0, 1]$.

### The HP $\Sigma$ Command

The HP lets you compute finite sums directly from the keyboard. For example, $2^2 + 3^2 + 4^2 + 5^2 + 6^2$ can be computed by successively entering the summation index, the lower and upper limits, the summand, and pressing the $\Sigma$ key:
'I' [ENTER] 2 [ENTER] 6 [ENTER] 'I^2' [ENTER] [⇒] [Σ]
to get the result, 90.

However, we will not want to carry out this procedure for every Riemann sum—we also need to do sums within a program! In *The Binomial Theorem and Calculus* project, we computed finite sums with programs that used the `FOR` loop construct. Here is a simpler code to compute $\Sigma_{j=m}^{n} j^2$ using the $\Sigma$ operation:

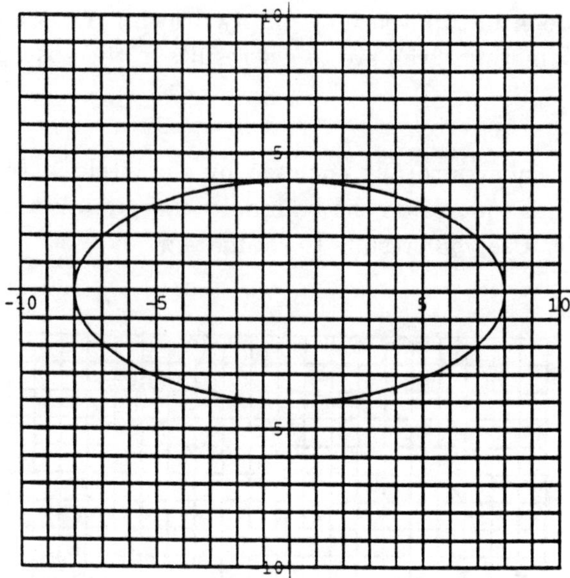


FIG. 11. Ellipse with superimposed grid.

≪ → M N 'Σ(J=M, N, J^2)' ≫
Here, the lower limit $m$ and the upper limit $n$ are read in from the stack. Try this code with $m = 2$ and $n = 6$ and check that you get 90 again.

**Exercise 5** Make sure you understand what Σ does by checking this result by hand.

## THE LAB

**Exercise 6** Use code like:
≪ → N 'Σ(J=1,N, J^3)' ≫
to evaluate sums of cubes for the cases $n = 4$, 16, 64, 256, and 1024 (i.e., $n = 4^k$, for $k = 1$, 2, 3, 4, and 5).

**Exercise 7** Write a summation code $\Sigma_{J=M}^{N} F(J)$ for an arbitrary `DEF`ined function `F(J)`. **Tip:** Always test new code with known results. For example, check your code with the data for the first part of the previous problem: `F(J)=J^3, M=1, N=4`.

### The Riemann Sum with the HP

We now use the HP Σ command to compute the Riemann sum based on the equal subdivisions and midpoints defined above in the Introduction. Here is the code:

```
@ RSUM--Compute Riemann sums (midpoint approximation)
@ Required globals:
@     F--DEFined function to be approximately integrated.
@ Local variables on stack:
@     A--left endpoint.
```

```
@     B--right endpoint.
@     N--number of subdivisions.
@ Other local variables:
@     DELX--length of subintervals.

RSUM:
<< -> A B N                                   @ read from stack
   << '(B-A)/N' EVAL -> DELX                  @ compute length of subintervals
      << 'sum(J=0,N-1,F(A+(J + 1/2)*DELX))' @ form sum; "sum"=right-shift-TAN
         EVAL DELX *                          @ evaluate and multiply by DELX
      >>                                      @ end scope of DELX
   >>                                         @ end scope of A, B, N
>>                                            @ end procedure
```

**Notes:** In the listing, we used the lower case word 'sum' to indicate the [⇒] [Σ] keystrokes that evoke the sum operation because of font limitations. Also, since $\Delta x$ = DELX is a constant, we took it outside the sum.

For the numerical evaluation of $\pi$, we've noted above that with the limits are $a = 0$, $b = 1$, the appropriate function is $f(x) = 4\sqrt{1 - x^2}$. DEFine this F and then the keystrokes 0 [ENTER] 0 [ENTER] 5 [RSUM] should yield the answer: 3.17199...

**Exercise 8** After repeating the above run as a check, determine $n$ to verify the approximation $\pi \approx 3.14$ (correct to three significant figures).

**Exercise 9** Approximate the area under one arch of the sine curve to three significant figures. That is, use $f(x) = \sin x$ and the interval $[0, \pi]$. **Note:** With calculus, the numerical value of this area can be determined *exactly*. In fact, the exact answer is 2. So if you aren't getting something around 2, take a deep breath and think about what you are doing.

**Exercise 10** Approximate the area under one arch of the curve $\sin(\sin x)$ to three significant figures. That is, use $f(x) = \sin(\sin x)$ and the interval $[0, \pi]$. **Note:** the numerical value of this area can*not* be determined *exactly*. It can only be approximated with methods such as the one you are using.

## AFTER THE LAB

**Exercise 11** Search your text for a formula that will allow you to check the results you obtained in Exercise 6.

**Exercise 12** The idea of approximating areas with curved boundaries by sums of simple areas is attributed to Archimedes. Find out when Archimedes lived and give at least three interesting accomplishments of his.

## BEFORE THE LAB

### Introduction

If the function $f$ has an explicit antiderivative $F$ in terms of elementary functions, then the Fundamental Theorem of Calculus tells us that

$$\int_a^b f(x)\,dx = F(b) - F(a),$$

so that evaluating the definite integral of $f$ can be achieved by merely evaluating $F$ at $a$ and $b$. For a continuous $f$, the Existence Theorem and Fundamental Theorem together guarantee that the definite integral and antiderivative exist (for example, $F(x) = \int_a^x f(t)\,dt$). However, for many elementary functions $f$, the antiderivative $F$ is not itself an elementary function that we can evaluate directly and hence the Fundamental Theorem does not provide a means for evaluation of the definite integral. In such cases, we can still get accurate numerical approximations to the definite integral—the Existence Theorem guarantees this! The numerical approximation of integrals is a major topic in its own right and we can only scratch the surface in a first course in calculus. In this project, we examine the first step—direct use of the Riemann sums that are used in the proof of the Existence Theorem.

In this and the next Project, we examine the *quality* of some basic numerical approximations, that is, we find out how good a job they do in estimating the exact integral. In this Project, you will see that some Riemann sums provide better approximations than others, even when using the same number of sub-intervals.

**Exercise 1** Use the Fundamental Theorem to check that

$$F(x) = \int_a^x f(t)\,dt$$

is an antiderivative of $f(x)$.

**Exercise 2** Show that the exact area between the $x$-axis and the function $f(x) = x^4 + 1$ on the interval $[0, 5]$ is 630.

**Exercise 3** Explain the geometrical meaning of the curve $y = \sqrt{100 - x^2}$ on the interval $[0, 10]$ and thus evaluate exactly the definite integral,

$$I = \int_0^{10} \sqrt{100 - x^2}\,dx.$$

## THE LAB

### The Riemann Sum with the HP

Start with the particular Riemann sum in which the *left* endpoint is used on each sub-interval. This is just a simple modification of the command used the *Area* project:

```
@ RSUM--Compute Riemann sums (left endpoint approximation)
@ Required globals:
@       F--DEFined function to be approximately integrated.
@ Local variables on stack:
@       A--left endpoint.
@       B--right endpoint.
@       N--number of subdivisions.
@ Other local variables:
@       DELX--length of subintervals.

RSUM:
<< -> A B N                              @ read from stack
   << '(B-A)/N' EVAL -> DELX             @ compute length of subintervals
      << 'sum(J=0,N-1,F(A + J*DELX))'    @ form sum; "sum"=right-shift-TAN
         EVAL DELX *                     @ Evaluate and multiply by DELX
      >>                                 @ end scope of DELX
   >>                                    @ end scope of A, B, N
>>                                       @ end procedure
```

**Exercise 4** As a "warm-up", we compute some Riemann sums for $f(x) = x^4 + 1$ on $[0, 5]$. Test your code with $n = 5$; you should get 359. Notice that compared to the exact answer (630), this is an awful approximation! Compute the Riemann sums for $n = 10$ and $n = 100$.

We now improve the above Riemann sum code, which is so far restricted to using the left endpoint of each sub-interval. Introduce a parameter $p$ (standing for "proportion"), where $p$ is a value between 0 and 1. Notice that when $p = 0$ the new `Sum` command below acts just like the above `RSUM` command. But when $p = 1$, we get the Riemann sum using the *right* endpoint of each sub-interval. The intermediate choice $p = 0.5$ gives the "midpoint rule" (as used in the Area project). Here is the improved code:

```
@ RSUM--Compute Riemann sums with equally spaced subintervals
@ Required globals:
@       F--DEFined function to be approximately integrated.
@ Local variables on stack:
@       A--left endpoint.
@       B--right endpoint.
@       N--number of subdivisions.
@       P--proportion, see text.
@ Other local variables:
@       DELX--length of subintervals.

RSUM:
<< -> A B N P                            @ read from stack
   << '(B-A)/N' EVAL -> DELX             @ compute length of subintervals
      << 'sum(J=0,N-1,F(A+(J+P)*DELX))'  @ form sum; "sum"=right-shift-TAN
         EVAL DELX *                     @ Evaluate and multiply by DELX
      >>                                 @ end scope of DELX
   >>                                    @ end scope of A, B, N, P
>>                                       @ end procedure
```

**Exercise 5** Repeat Exercise 4, this time using *right*-endpoint sums. Contrast the results.

**Exercise 6** Make your study systematic by filling in an *error* table with column headings like this:

```
left-endpoint (p = 0)     midpoint (p = 0.5)      right-endpoint (p = 1)
```

and rows labeled by $n$ =10, 100, and 1000. The nine entries in the table should be the errors in the corresponding Riemann sum. (It is not necessary to do anything fancy unless you want to, just get the nine error values and write them in a table.)

**Exercise 7** Do the analog of the previous problem for the function and interval in Exercise 3.

## AFTER THE LAB

**Exercise 8** The errors in numerical approximations tend to be proportional to $1/n$, or to $1/n^2$, or to $1/n^3$ and so on, where $n$ is the number of sub-intervals. Try to detect this pattern: on the basis of the tables you made in Exercises 6 and 7, find the power of $n$ for each of the values of $p$. **Hint**: When $n$ increases by a factor of 10 (say from 10 to 100), then if the error decreases by a factor of 10 it is behaving like $1/n$. On the other hand, if the error decreases by a factor of 100, then it is behaving like $1/n^2$, etc. Fractional powers are possible.

**Exercise 9** Can you give an intuitive explanation of why Riemann sums with $p = 1/2$ should usually have smaller errors than those with $p = 0$ or 1?

**Exercise 10** Let

$$f(x) = e^{\sin\sqrt{x^{12}+\cos^2 x}}.$$

Does $\int_0^1 f(x)\,dx$ exist? Does the Fundamental Theorem help you to evaluate it? If so, how? Could you get a good numerical approximation? If so, how?

## BEFORE THE LAB

This project reviews the connection between the area function and the Fundamental Theorem and then implements three basic numerical integration methods with the HP: the trapezoid rule, the midpoint rule and Simpson's rule.

### The Area Function

We study the area function $A(x)$ associated with a continuous function $f(x)$. $A(x)$ is defined as the integral of $f$ from a given, fixed left endpoint $a$ to the variable point $x$:

$$A(x) = \int_a^x f(t)\,dt.$$

**Exercise 1** Using geometry (i.e, no calculus allowed!), construct the area function for the following functions:

a) $f(x) = c,\qquad c = \text{constant}$

b) $f(x) = cx,\qquad c = \text{constant}$

**Exercise 2** Explain why $A'(x) = f(x)$.

**Exercise 3** If $f(x) = 5$ and $a = 1$, show that $A(x) = 5x - 5$.

### The Basic Numerical Integration Methods with the HP

In a previous project, we gave the following code for computing Riemann sums:

```
@ RSUM--Compute Riemann sums with equally spaced subintervals
@ Required globals:
@      F--DEFined function to be approximately integrated.
@ Local variables on stack:
@      A--left endpoint.
@      B--right endpoint.
@      N--number of subdivisions.
@      P--proportion, see text.
@ Other local variables:
@      DELX--length of subintervals.

RSUM:
<< -> A B N P                               @ read from stack
   << '(B-A)/N' EVAL -> DELX                @ compute length of subintervals
      << 'sum(J=0,N-1,F(A+(J+P)*DELX))'     @ form sum; "sum"=right-shift-TAN
         EVAL DELX *                        @ Evaluate and multiply by DELX
      >>                                    @ end scope of DELX
   >>                                       @ end scope of A, B, N, P
>>                                          @ end procedure
```

The parameter $p$ controls the kind of Riemann sum formed: $p = 0$ gives the Left Rectangle Rule, $p = 1$ gives the Right Rectangle Rule and $p = 0.5$ gives the Midpoint Rule. Valid, but less popular, Riemann sums are given for other choices of $p$ between 0 and 1. Assuming that you have DEFined 'F(X)=SIN(X), you use RSUM like this to get the midpoint sum ($p = 0.5$) for $a = 0$, $b = \pi$, and $n = 5$:

0 [ENTER] [⇐] [π] [⇐] [→ NUM] 5 [ENTER] .5 [RSUM]

You should get the result 2.03328.... In this project, we compare three basic numerical integration methods (by the way, there are entire texts devoted solely to developing and analyzing numerical integration methods!). The first is the *Midpoint rule*. This is simply RSUM with $p = 0.5$, so we obtain the procedure:

```
MIDP:  @ MIDPOINT RULE BASED on RSUM
<< -> A B N
   << A B N .5 RSUM   @ RSUM with p = 1/2
   >>
>>
```

The Trapezoid rule can be defined as the average of the Rsums with $p = 0$ and $p = 1$ (Left and Right Rectangle rules), so

```
TRAP:  @ TRAPEZOID RULE BASED on RSUM
<< -> A B N
   << A B N 0 RSUM   @ left endpoint sum
      A B N 1 RSUM   @ right endpoint sum
      + 2 /          @ average left and right
   >>
>>
```

Finally, Simpson's rule can be defined as the weighted average of the Midpoint rule and the Trapezoid rule with the former having twice the weight of the latter, so that

```
SIMP:  @ SIMPSON'S RULE BASED on TRAP and MIDP
<< -> A B N
   << A B N TRAP      @ Trapezoid rule
      A B N MIDP 2 *  @ Midpoint rule (gets double weight)
      + 3 /           @ weighted average
   >>
>>
```

**Check:** With $f(x) = \sin x$, $a = 0$, $b = \pi$, and $n = 5$, SIMP should give the value 2.00011 to 6 significant figures (don't forget to convert $\pi$ to a numerical value as above).

For easy reference, we give the explicit formulas with error term for the three rules examined in this project. With the usual $\Delta x = (b-a)/n$, we use the notation, $f_i \equiv f(a + i\Delta x)$, to write:

$$\begin{aligned}
\text{Mid} &: \Delta x(f_{\frac{1}{2}} + f_{\frac{3}{2}} + \cdots + f_{n-\frac{1}{2}}) - \frac{f''(\theta_M)(b-a)^3}{24n^2} \\
\text{Trap} &: \Delta x(\frac{f_0}{2} + f_1 + f_2 + \cdots + f_{n-1} + \frac{f_n}{2}) + \frac{f''(\theta_T)(b-a)^3}{12n^2} \qquad (5) \\
\text{Simp} &: \frac{\Delta x}{3}(f_0 + 4f_1 + 2f_2 + 4f_3 + \cdots + 2f_{n-2} + 4f_{n-1} + f_n) + \frac{f^{(4)}(\theta_S)(b-a)^5}{180n^4}.
\end{aligned}$$

In each case, the final derivative term represents the error in that rule and the $\theta$'s denote unknown points in the interval $[a, b]$, just as in the mean value theorem.

**Exercise 4** It comes as a bit of a surprise that Simpson's Rule gives the *exact* result for cubic polynomials. Use the theory expressed in Equation 5 to explain this.

## THE LAB

**Exercise 5** The integrals of $(k+1)x^k$, $k = 1, 2, 3, \ldots$ from $x = 0$ to $x = 1$ are all equal to one. Compare the results obtained by the Midpoint, Trapezoid and Simpson rules for $1 \leq k \leq 5$ and $n = 5$ subdivisions. **Tip**: Instead of changing F five times, use code like this:

```
EX5:
<< CLEAR                             @ empty stack
   1 5 FOR K                         @ 5 powers of X
     'F(X)=(K+1)*X^K' DEFINE         @ DEFine function
     K "K" ->TAG                     @ print labeled K value
     0 1 5 MIDP
     0 1 5 TRAP
     0 1 5 SIMP
   NEXT
   'F' PURGE                         @ clean up
>>
```

The "TAG" statement is available from the keyboard by: [PRG] [TYPE] [→TAG]. You can examine the results by using [△]—the stack environment—to move up to the top of the stack.

**Exercise 6** With the function $f(x) = \sin x$ and interval $[0, \pi]$, we study the errors obtained by the Midpoint, Trapezoid and Simpson rules as a function of the number of sub-intervals, $n$. To make the study easier, we provide you a loop that computes these *errors*. These errors are called `MidE`, etc. Note that $n$ increases as powers of 2 (as $k = 1, 2, \ldots, 7$ in the `Do` loop). This will make it easy for you to see the "order of convergence" of the various rules.

```
EX6:
<< CLEAR                             @ empty stack
   'F(X)=SIN(X) DEFINE               @ DEFine function
   2 -> EXACT                        @ integral equals 2
   << 1 7 FOR K                      @ 7 powers of two
         '2^K' EVAL 'N' STO          @ compute N = 2^K
         N "N" ->TAG                 @ print labeled N value
         EXACT                       @ Put EXACT answer on stack
         0 pi ->NUM N MIDP           @ get midpoint approximation
         - "MidE " ->TAG             @ subtract to get error and print
         EXACT
         0 pi ->NUM N TRAP           @ same for Trapezoid rule
         -  "TrapE" ->TAG
         EXACT
         0 pi ->NUM N SIMP           @ and for Simpson's rule
         -  "SimpE" ->TAG
```

```
      NEXT
    >>
    'F' PURGE                  @ clean up globals
    'N' PURGE
>>
```

**Exercise 7** Repeat the last exercise with $f(x) = \sin(\sin(x))$. There is no elementary function antiderivative for this function, so you cannot get an exact answer and numerical approximations are required. The HP's built-in $\int$ command computes most definite integrals accurately. So, to get a good approximation to the exact value, enter the [⇒] [SYMBOLIC] [INTEGRATE] environment and [CHOOS] the NUMERIC mode after setting the function and limits. Note that after performing a numerical integration, the global variable IERR holds an estimate of the integration error.

**Exercise 8** Do it one more time with $f(x) = \sqrt{4 - x^2}$ and with $a = 0$ and $b = 2$. For this function, you can get the exact answer from knowledge of the area of a circle.

## AFTER THE LAB

**Exercise 9** In regard to the area function $A(x)$ defined above:

a) What is $\dfrac{dA(x)}{dx}$?

b) What is the value of $A(a)$, where $a$ is the left endpoint? Why?

c) Corresponding to a continuous function $f(x)$, suppose that a clever friend has found an antiderivative, call it $F(x)$. How is her $F(x)$ related to $A(x)$?

**Exercise 10** The errors in numerical integration generally behave like $C/n^k$. From your results for Exercise 6, determine $k$ and $C$ for each of the three integration rules examined there. Also, note the relation between the errors in the Trapezoid and Midpoint rules. Answer carefully: do your results agree with Equation 5?

**Exercise 11** Repeat the last exercise using the results you obtained in Exercise 7.

**Exercise 12** And repeat again using the results you obtained in Exercise 8.

## BEFORE THE LAB

**Exercise 1** We know that the equation $x^2 + y^2 = r^2$ represents a circle of radius $r$. Use this fact to justify the formula:

$$A_{circle} = 4r \int_0^r \sqrt{1 - (x/r)^2}\, dx = \pi r^2 \tag{6}$$

for the area of this circle.

**Exercise 2** Similarly, the equation

$$\left(\frac{x}{a}\right)^2 + \left(\frac{y}{b}\right)^2 = 1$$

represents an ellipse with axes $a$ and $b$. Justify both of the formulas:

$$A_{ellipse} = 4b \int_0^a \sqrt{1 - (x/a)^2}\, dx \tag{7}$$

$$= 4a \int_0^b \sqrt{1 - (y/b)^2}\, dy \tag{8}$$

for the area of this ellipse.

**Exercise 3** Make the substitution $u = x/a$ in Equation 7 and use Equation 6 to compute $A_{ellipse}$ explicitly. Check your result in each of the following two ways: (1) derive the result again using Equation 8 and (2) show that your result specializes correctly in the case when the ellipse becomes a circle of radius $r$.

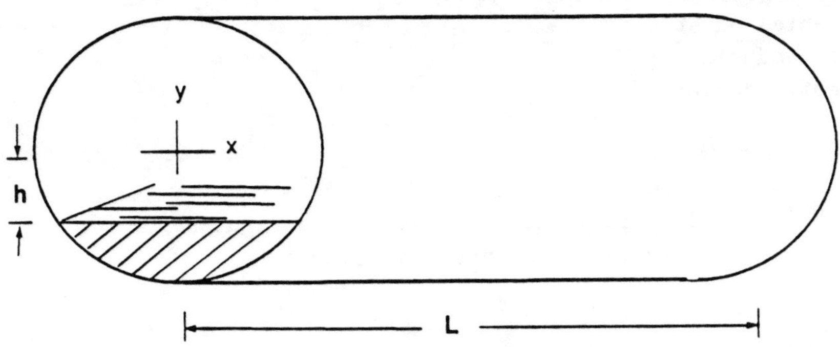


FIG. 12. Partially filled tank.

### The Tank Problem

Figure 12 shows an underground fuel tank (e.g., as at a service station) in the shape of an elliptical cylinder. The tank is lying on its side (i.e., the cylinder is horizontal) and has length $L$ ft. The elliptical cross-section has axes $a$ ft and $b$ ft. The owner has only a crude way of measuring the contents: he lowers a stick down into the tank and measures how high the fuel level is on the stick. You are to help him convert this measurement, call it $h$ (in ft), into cubic ft of fuel in the tank.

**Exercise 4** Derive the following formula for the volume $V = V(h)$, where $-b \le h \le b$:

$$V = 2La \int_{-b}^{h} \sqrt{1-(y/b)^2}\, dy \tag{9}$$

**Hint**: the volume is an area times the length of the tank.

## THE LAB

### The HP $\int$ Command

In the *Numerical Integration II* project, you used the [SYMBOLIC] [INTEGRATE] environment to interactively compute definite integrals.

The HP also lets you compute definite integrals directly from the keyboard. For example, $\int_0^2 x^3\,dx$ can be computed as:
0 [ENTER] 2 [ENTER] 'X^3' [ENTER] 'X' [ENTER] [⇒] [$\int$] [⇐] [→NUM]
to get the result, 4.

Here is a simple code to conveniently compute $\int_a^b f(x)\,dx$ using the $\int$ operation:
≪ → A B '∫(A, B, F(X))' →NUM ≫
In the listing below, which repeats this code in our usual format, the phrase "int" is used to indicate the $\int$ key (right shifted [COS]):

```
@ DEF--Compute numerical approximation to a definite integral
@ Required globals:
@       F--DEFined function of X to be numerically integrated.
@ Stored variables:
@       IERR--integration error
@ Local variables on stack:
@       A--left endpoint.
@       B--right endpoint.

DEF:
<< -> A B               @ read from stack
   'int(A B F(X),X)'    @ do integration; "int" means right-shift-COS
    ->NUM               @ evaluate in numerical mode
>>
```

DEFine the function 'F(X)=X^3' and try this code with $a = 0$ and $b = 2$ and make sure that you get 4 again.

Tip: Definite integration of complicated functions can take a long time. Numerical integrations on the HP can be speeded up by using the MODE environment to set your number display format to a lower accuracy factor (say [FIX 3]). This decreases the accuracy from the highly accurate (and very slow) [STD] mode. Please change MODEs and then repeat the use of DEF on $f(x) = x^3$ with $a = 0$ and $b = 2$ one more time.

**Exercise 5** With $r = 1$, use the DEF procedure to numerically check Equation 6. Describe your work in detail. **Remark**: Don't forget to change F(X).

**Exercise 6** With $a = 1$ and $b$=2. use the HP to check your evaluation of the integrals in Equations 7 and 8. Describe your work in detail.

**Exercise 7** Continuing with the tank problem: using Equation 9 with the specific parameter values, $L = 20$. $a = 10$ and $b = 5$. compute $V(h)$ from $h = -5$ to $h = 5$ with unit spacing. Coding hint:

```
<< -5 5 FOR K
     -5 K DEF
   NEXT
>>
```

**Exercise 8** Your friend would like to put a mark on the stick as a warning to indicate when only 500 $\text{ft}^3$ are left. Using the same numerical parameter values as in Exercise 7, figure out the height $h$ at which the stick should be marked. **Hints:** Use Newton's method to solve the equation $V(h) = 500$. Newton's method tries to find a solution to $f(x) = 0$, what is the $f$ for this problem? Newton's method requires a starting guess, how can you get a good one? Newton's method requires a derivative—in this problem the derivative is easy to get, why? Finally, express your answer as the distance from the *bottom* of the stick. Coding hint:

```
...
-5 H DEF 500 -    @ compute V(H) from F
...               @ compute V'(H)
...
```

## AFTER THE LAB

**Exercise 9** Show that the total volume of the cylindrical tank is $\pi abL$ and use this result to check your output in Exercise 7 for the values $h = -5$, $h = 0$, and $h = 5$.

## BEFORE THE LAB

### Introduction

As you know, if a constant force $F$ is applied over a distance $d$ the work done is simply the product $Fd$. More frequently the force is changing due to any number of things (e.g., the terrain, the wind, a magnetic field, etc.). Moreover, often the force is not applied in a single direction, and this complicates things a bit. You will be exploring this important notion in a simple setting in which we assume:

a) The force is a function only of the horizontal variable $x$ (i.e., $F = F(x)$).

b) The force is exerted directly along the curve of motion.

c) The curve of motion is a planar curve explicitly defined by $y = h(x)$ on some interval $[a, b]$.

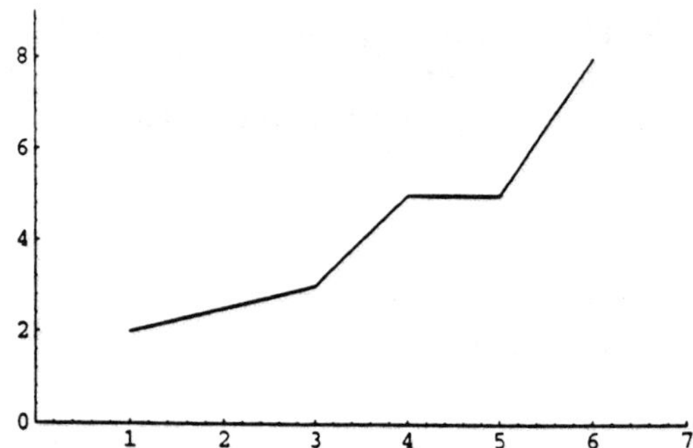


FIG. 13. The piecewise linear curve $C_1$.

**Exercise 1** Without using calculus,

a) Calculate the length of the curve $C_1$, the piecewise linear curve connecting the points, $(1, 2)$, $(3, 3)$, $(4, 5)$, $(5, 5)$, and $(6, 8)$, shown in Figure 13.

b) Calculate the length of the curve $C_2$, defined by: $y = \sqrt{4 - x^2}$, for $-2 \leq x \leq 2$. **Hint:** Just use some geometry.

c) By approximating with a few inscribed chords, estimate the length of the curve $C_3$, defined by: $y = x^2 - 4$, for $0 \leq x \leq 2$.

**Exercise 2** In this problem, $k$ is a constant, $m(x)$ is the slope of the curves $C_1$, $C_2$, and $C_3$ defined in Exercise 1. Find $m(x)$ for each of these curves and then without doing any integrals,

a) Calculate the work done if the force $F = k \cdot m(x)$ is applied along the curve $C_1$.

b) Calculate the work done if the force $F = k$ is applied along the curve $C_2$.

c) Estimate the work done if the force $F = k \cdot m(x)$ is applied along the curve $C_3$.

**Exercise 3** Now let $C$ represent a general planar curve defined by $y = h(x)$ on the interval $[a, b]$ and let $F(x)$ be a force always exerted in the direction of $C$.

a) By considering the element of work done on a small interval $[x, x + dx]$, carefully explain the following formula:

$$\text{Work} = \int_a^b F(x)\sqrt{1 + (h'(x))^2}\,dx$$

b) Use this formula to compute the work done by $F(x) = 3x$ along the curve $C_3$ defined above.

## THE LAB

Begin by setting up HP procedures to compute the length and work for the situation described in Exercise 3. These routines assume you've STOred the expression H(X) for the curve and F(X) for the force function.

```
@ DS--Compute differential arclength
@ Required globals:
@      H--STOred expression for H(X).
DS:
<<  'X' PURGE     @ clear out any previous X
    H 'X' deriv   @ get H'; "deriv" means right-shift-SIN
    SQ 1 + sqrt   @ ds = sqrt(1 + h'^2); "sqrt" means the square root key
    'DSDX' STO    @ store in DSDX
>>

@ ARCL--Compute arclength
@ Required globals:
@      DSDX--STOred expression for ds (use DS).
@ Local variables on stack:
@      A, B--endpoints.
ARCL:
<< -> A B
   << A B DSDX 'X' @ set up integration arguments
      int ->NUM    @ get numeric arclength; "int" means right-shift-COS
   >>
>>

@ WORK--Compute work
@ Required globals:
@      F--STOred expression for force F
@      DSDX--STOred expression for ds (use DS).
@ Local variables on stack:
@      A, B--endpoints.
WORK:
<< -> A B
   A B 'F*DSDX' EVAL 'X' @ set up integration arguments
   int ->NUM             @ get numeric work; "int" means right-shift-COS
   >>
>>
```

**Exercise 4** Using the HP, compute the lengths of $C_2$ and $C_3$. Compare with your earlier results.

**Exercise 5** Using the HP, compute the work associated with each of the following curves and forces:

a) Check your result for the second part of Exercise 3.

b) Let $C$ be defined by

$$h(x) = \frac{\sin 3x}{1 + x^2}$$

on the interval $[0, \pi]$ and let $F(x) = 2x^2$. Compute both the length and work.

c) Compute the work for the last curve, but now with a force equal 3 times the slope of the curve.

## AFTER THE LAB

**Exercise 6** Considering the length of the curve and the magnitude of the force in the last part of Exercise 5, the work is rather small. Can you explain this?

**Exercise 7** The second part of Exercise 3 was designed so that you could evaluate the work integral "by hand." Construct a *new* example of an $h$ and $F$, so that you can similarly evaluate the work integral. **Challenge:** Construct a whole *family* of $h$'s and $F$'s that lead to easy work integrals.

## BEFORE THE LAB

### Introduction

An important problem in science and engineering is to find a functional relationship between two (or more) variables. For example, in some physical setting:

a) How are temperature and pressure related?

b) How are velocity and time related?

Often this question is addressed by examining a discrete set of data—e.g., one gathered by running several, or many, experiments. In this project you will be trying to establish an approximate linear (or quadratic) relationship from sets of data.

**Exercise 1** Consider the following set of five $x$-values. followed by the $y$-values:
$x$-values: .2. 1.1, 2.0. 3.1, 3.9, 5.1
$y$-values: -2, 0, 1, 2.8, 5, 7

First carefully plot the five $(x, y)$ points on a piece of graph paper. You will notice that the points almost, but not quite, lie on a line. Assuming that these points represented an approximation to a linear relation, attempt to find this relation by drawing a line on your graph that seems to best "fit" this data. Write down the equation of your line.

We now see how the problem of attempting to fit a line, or another curve, through a set of discrete data can be done using the HP. Let's put the data from Exercise 1 into the HP. To do this, enter the `Statistics` environment with ⇒ STAT and select `Fit data...`. Notice that the default model is `Linear Fit` which is exactly what we want, so we need only enter our data. To do this, select the `EDIT` menu item. This brings up a special form of the `MATRIX` writer. As you ENTER each $x$-value, the entry point drops to the next line, after entering the sixth and last $x$-value. press ▷ and the entry point moves to the top of the second column. Then key in each $y$-value followed by ENTER. When all the data is inserted, select the model field (which reads `Linear Fit`) and use `OK`. The stack will show three fields—level one shows the covariance of $x$ and $y$. We won't dwell on this one; go ahead and press DROP to get rid of it. The next field gives the correlation of the fitted $y$-values to the actual $y$-values—highly correlated data has correlations near $\pm 1$ and very weakly correlated data has correlations near 0. After noting that the correlation is very high (as expected, a linear fit does a good job with this data set), you can DROP that one too. This leaves the fitted straight line, $y = -2.36 + 1.82x$, on the stack (actually, the stack contains an expression representing the right side of this line).

**Remark:** Data can be highly correlation (say .96) and yet the fit can be poor: just because high $y$-data values are "highly correlated" with high $y$-fitted values and vice versa, doesn't mean the details of the fit are satisfactory. Using correlation as a measure of "goodness of fit" doesn't make sense unless the correlation is very high (e.g., .99).

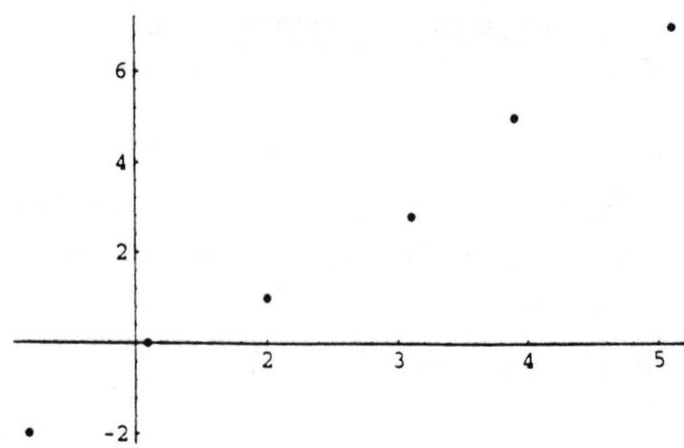


FIG. 14. Plot of the data.

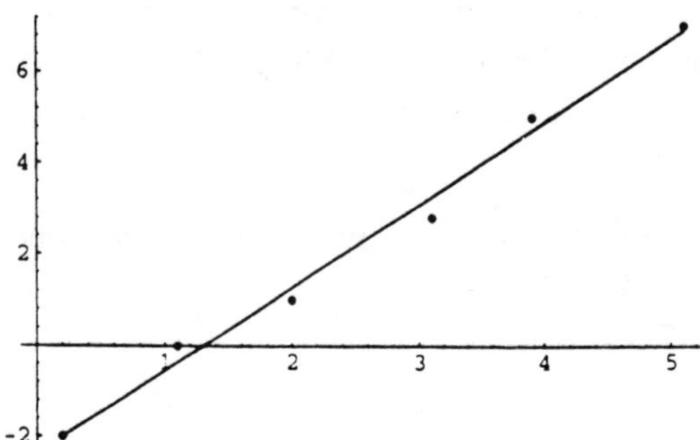


FIG. 15. Plot of the data with best fitting line.

A scatter plot of the data can be obtained by pressing [⇐] [STAT] [PLOT] [SCATR] (like Figure 14, but, as usual, without the axes labeling). After the plot is drawn press [STATL] to superimpose the fitted straight line (akin to Figure 15).

**Note**: This line fits this set of points "best" in the sense of "least squares" (in the errors). This will be a recurring subject in your career. This process is an example of "curve fitting" a discrete set of data, or, as said above, seeking a functional relation between $x$ and $y$.

## THE LAB

**Exercise 2** The data below came from the Allegheny National Forest in Pennsylvania. The issue addressed was: can either the diameter or the height of a tree accurately predict the volume of wood in the tree? Using the data provided below, you are to see if there is a meaningful (linear) relation between:

a) diameter and volume, or

b) height and volume.

Below, the "ddata" represents the diameter data for 11 different trees and "hdata" is the height data and "vdata" is the volume data for the same trees. Which of the two (diameter or height) would be the better estimator of volume? Roughly what volume would you expect from a tree of diameter of 19 inches? How much error would you expect in this estimate? Why? Turn in two key sketches supporting your work. **Remark**: Think about whether **ddata**, etc., is analogous to **xdata** or to **ydata** in the above example.

```
ddata = {8.6, 10.7, 11., 11.4, 12.0, 13.3, 14.5, 16.0, 17.3, 18.0, 20.6}
hdata = {65, 81, 75, 76, 75, 86, 74, 72, 81, 80, 87}
vdata = {10.3, 18.8, 18.2, 21.4, 19.1, 27.4, 36.3, 38.3, 55.4, 51.5, 77.}
```

**Tip:** You can delete old data with DEL OK. Also, you can modify data by moving the cursor to the desired place and entering the new data—in particular, you don't have to retype the **vdata** in the two parts of this problem. Just type a final ENTER when you are satisfied with the keyed in data.

**Exercise 3** The following data came from velocity readings from a Porche 914 acceleration test. Try a linear fit and report your results including a sketch of the data and fitted line. First the time data, then the velocity data:

```
tdata = {0, 1.8, 2.8, 3.9, 5.1, 6.8, 8.6, 10.5, 13, 16.3, 19.7, 25.6}
vdata = {0, 30, 40, 50, 60, 70, 80, 90, 100, 110, 120, 130}
```

**Exercise 4** In the previous problem, you will have seen that a linear fit is not very good despite a high correlation. By looking at the shape of the data, decide on another function that you think would give a better fit. The HP does not allow you to pick any fitting function you'd like, but it does have the most common choices:

**Linear** $y = b + mx$

**Logarithmic** $y = b + m \ln x$

**Exponential** $y = be^{mx}$

**Power** $y = bx^m$

These choices are available by highlighting the MODEL field and CHOOSing the desired model. Pick the model that includes the function that you decided would give a better fit. Report whether the correlation is significantly improved with this model and whether the fitting parameters are in accord with your idea of a good fitting function.

## AFTER THE LAB

Here we introduce the "method of least squares" which is the technique used by the HP in implementing the `Fit` command. In our application of this important approximation technique, we have a set of $n$ data points, $(x_i, y_i)$ and want to find the line, $y = ax + b$, that best fits the $n$ data points in the sense that the sum of the squares of the errors in this approximation is as small as possible. This is clearly a minimization problem and the function to be minimized is

$$F(a, b) = \sum_{i=1}^{n} [(ax_i + b) - y_i]^2.$$

Notice that despite all the $x$'s and $y$'s lying around, the *unknowns* are $a$ and $b$—we want to find the $a$ and $b$ that minimize $F(a, b)$ for the *given* $x_i$'s and $y_i$'s.

Since we don't yet know how to minimize functions of more than one variable, we simplify the problem by assuming that we know the value of $b$, say, $b = 0$. Hence we seek the best line of the form $y = ax$. In this case $F = F(a)$ and we have a one-variable minimization problem that we *do* know how to handle.

**Exercise 5**

a) Assuming $b = 0$, show that in seeking the minimum of $F(a)$, one is led to the value:

$$a = \frac{\sum_{i=1}^{n} x_i y_i}{\sum_{i=1}^{n} x_i^2}.$$

b) Still assuming $b = 0$, find, by "hand," the least square line through the data: (1, 1), (2, 4), (3, 5), (4, 6).

c) On graph paper show the resulting line and the data points. Was the assumption that $b = 0$ supported by this particular data set?

**Exercise 6** In Exercise 2, you may have been surprised that the attempt to find that a linear relation between diameter and volume of the trees was as successful as it was. What relation would you *expect* to find (think about the shape of a tree)?

**Exercise 7** Again in reference to Exercise 2, speculate on why the data did not support finding a linear relation between height and volume of the trees.

## BEFORE THE LAB

### Introduction

We continue the theme of curve fitting a set of discrete data. However, in this project, the functional relationship in the data is a little more complex. Consequently, either the $x$-data or the $y$-data. or both, need to be "transformed." Typically, one has theory supporting a certain functional relationship between $x$ and $y$, or such a relationship is suggested by the data itself. In this project, we study cases in which a logarithmic transformation of the data produces a linear relationship in the new (transformed) variables.

**Exercise 1** Suppose we know, either from theory or empirically, that the power law relationship (or "model")

$$y = k\,x^m$$

holds between $x$ and $y$. By taking the logarithm of this equation, show that one obtains the *linear* relationship,

$$Y = K + mX.$$

where $X = \ln x$, $Y = \ln y$, and $K = \ln k$ (so that we compute the model parameter $k$ as $k = e^K$).

**Exercise 2** Now consider the exponential model

$$y = c\,e^{ax}.$$

As in Exercise 1, obtain the linear equation

$$Y = C + aX$$

and identify $X, Y$ and $C$ in terms of the original model quantities. How would you compute the model parameters $c$ and $a$ from the fit to this linear equation?

**Exercise 3** Now consider the "logistic" population model,

$$y = \frac{M}{1 + e^{c-ax}}, \qquad M > 0. \tag{10}$$

Show that if $y(0) < M$, then $y(x)$ is strictly increasing for $x > 0$, and also show that $\lim_{x\to\infty} y = M$. To linearize the logistic relationship, temporarily think of $c - ax$ as the unknown and derive the equation $Y = c - ax$. What is $Y$ in this equation?

**Remark:** In population models, $M$ is the maximum sustainable population taking into account the restrictions in space, food, etc.

### Illustrative Example

We revisit the tree data problem from the previous project on *Curve Fitting for Discrete Data Sets*. If necessary, review that project to clear the statistics array and re-enter the data for the ordinary linear fit:

```
ddata = {8.6, 10.7, 11., 11.4, 12.0, 13.3, 14.5, 16.0, 17.3, 18.0, 20.6};
vdata = {10.3, 18.8, 18.2, 21.4, 19.1, 27.4, 36.3, 38.3, 55.4, 51.5, 77.};
```

Continuing the review, go ahead and reproduce the old linear fit (see Figure 16):

```
-41.1313 + 5.38555 x
```

Note that the correlation of this fit is .977. **Tip:** The User's Guide (Edition 7), page 21-2,

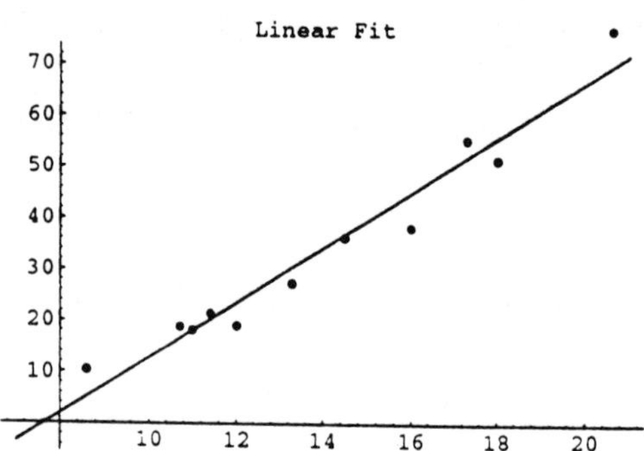


FIG. 16. Linear fit to the tree data.

tells how *To store the array in ΣDAT in a different variable*. Read this section and store the above data in the variable `'TREE'`. (This will save you typing it again later.)

As in Exercise 1, we compute the logarithm of the data. the HP "knows" how to apply functions like the natural log ([⇒] [LN]) to lists of data, so we can get the logs we need by transforming our 11 row by 2 column array to a list of length 22, taking the log, and then transforming back to an array. Here's one way to do it:

[⇒] [STAT] [Fit data...]
[NXT] [CALC] (puts array at level 1 of stack)
[PRG] [TYPE] [OBJ→] (split up the array)
[DROP] 22 [ENTER] (change list of array dimensions to number of items)
[→LIST] [⇐] [LN] (convert to list and do log transform)
[OBJ→] (split up list)
[DROP] [⇐] [{ }] 11 [SPC] 2 [ENTER] (change number of items to list of array dimensions)
[→ARR] (create array of log values)
[⇐] [CONT] [OK] (put array back into the `FIT DATA` environment)

You will now have data like this (perhaps with a different number of significant figures displayed):

```
{{2.15176, 2.33214}, {2.37024, 2.93386}, {2.3979, 2.90142},
 {2.43361, 3.06339}, {2.48491, 2.94969}, {2.58776, 3.31054},
 {2.67415, 3.59182}, {2.77259, 3.64545}, {2.85071, 4.01458},
 {2.89037, 3.94158}, {3.02529, 4.34381}}
```

Now do a linear fit to the logs of the data, to obtain:

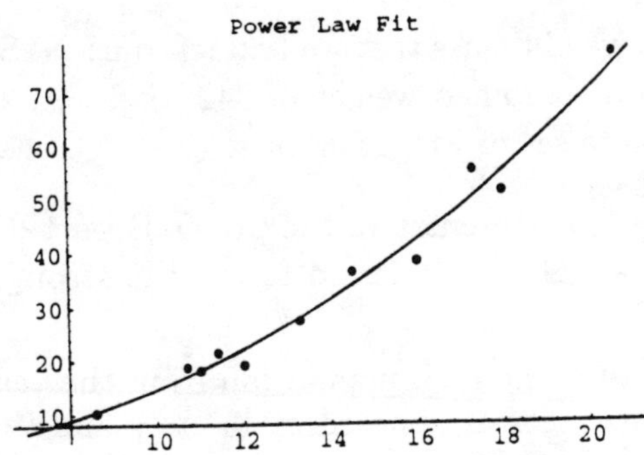


FIG. 17. Power law fit to the tree data.

```
-2.46854 + 2.24106 x
```

with correlation .991. According to Exercise 1, the power law fit, $k\,x^m$, directly uses the $m$-value (2.24106) and uses the $K$-value (-2.46854) to get $k = e^{-2.46854} = 8.4708210^{-2}$ (cf. Exercise 1). A plot of the power law fit, $k\,x^m$, vs. the original data is shown in Figure 17. In accordance with the change in correlation of the original linear fit from .977 to .991, our eye confirms that the improvement over the linear fit is substantial.

Now re-enter the **Fit data** environment and CHOOS the **TREE** data you previously saved (or type in again, if you didn't save it). Then highlight the **MODEL** field and CHOOS the **Power Fit** model. Notice that the result of the power fit is exactly what we obtained above—the HP builds in our power fit theory! Similarly, the HP builds in the **Exponential Fit** theory derived above and also two other fits: **Logarithmic Fit** and **Best Fit**. The latter chooses the fit that has the highest correlation from among the other four choices provided by the HP.

## THE LAB

**Exercise 4** Below is data relating the period of revolution $T$ (in days) of the 6 inner planets and their semi-major axes $a$ (in $10^6$ km). Use the HP's built in **Power Fit** to find $k$ and $m$ in Kepler's empirical power law relationship, $T = ka^m$. Hand a sketch showing the effectiveness of your fit. Newton was later able to justify the theoretical value $m = 3/2$ by assuming an inverse square force law for gravitational attraction—how well does the empirical data bear out Newton's theory?

```
Tdata = {87.97, 224.70, 365.26, 686.98, 4332.59, 10759.20}
adata = {58, 108, 149, 228, 778, 1426}
```

**Exercise 5** Below is data pertaining to the 1988 Olympic weight-lifting competition. The **xdata** is the "class data" giving seven weight classes from Featherweight to Heavyweight-2 (given in kg). The **wdata** is the combined weight lifted by the winners in each class.

```
xdata = {56, 67.5, 75, 82.5, 90, 100, 110};
wdata = {292.5, 340, 375, 377.5, 412.5, 425, 455}
```

a) Seek to fit this data in the form $w = kx^m$ as in Exercise 4. Find $k$ and $m$. (Note: theoretically, one should get $m = 2/3$. More on this later).

b) Add to the data one weight class that we had left out: N. Suleymanoglu of Turkey won the 60 kg class with a combined weight of 342.5 kg. Repeat the work of the previous part with the complete set of weight classes.

c) Comment on what the extra entry in the previous part did to your experiment. And explain why Suleymanoglu was referred to as "the strongest man in the world."

**Exercise 6** Below is data for the U.S. population from the years 1800 to 1990. The years are in the **xdata** array (year 1800 is denoted by 0, etc.) and the population, in millions, is given in the **ydata** array. First show that the popular exponential growth law, $y = c\exp ax$ does not work very well. Illustrate this failure with a graph. Then, as in Exercise 3, seek the relationship $y = M/(1+\exp(c-ax))$ and linearize to get $Y = c - ax$. Here you will need to make an "educated guess" at $M$, the maximum population attainable. For example, is $M = 300$ (representing 300,000,000) a good choice or will a smaller or larger $M$ make a better model? You are to investigate this, trying to find an $M$ that makes the model work well. (There is no perfect $M$). In particular, find an $M$ that you think will work well in estimating the population in the year 2050 ($x = 2050 - 1800 = 250$). What is your resulting estimate?

```
xdata = {0, 20, 40, 60, 80, 100, 120, 140, 160, 170, 180, 190}
ydata = {5.3, 9.638, 17.07, 31.44, 50.19, 76.2, 106.0, 132.2, 179.3, 203.3,
         226.5, 248.7}
```

**Tips:** Since you need to try several $M$ values, `STO`re the original census data in variable. After choosing $M$, you need to transform the $y$-column according to $Ydata = \ln(M/ydata - 1.0)$. The procedure for this transformation is similar to that used above in transforming the entire data array and is given in detail at the top of page 21-5 of the User Guide. If the transformation had involved any + signs, you'd have to replace those with [MTH] [LIST] [ADD]. Fortunately, this is not a concern with the current transformation (eg., [$\frac{1}{x}$] 300 [×] 1 [−] [⇒] [LN]), but this "glitch" always has to be kept in mind when applying functions to lists. Note that after the transformation, you will use `Linear Fit`. To judge the quality of the fit, you can use the correlation together with a plot of the data with linear fit (see the previous *Curve Fitting for Discrete Data Sets* project). To get the predicted value, use the `Fit Data` [PRED] menu item, with a 250 in the `X` field. You then have to transform this *linear* predication back to the original data domain (e.g., by highlighting the predicted value and using [CALC] [⇐] [$e^x$] 1 [+] [$\frac{1}{x}$] 300 [×]).

## AFTER THE LAB

**Exercise 7** While doing Exercise 6, you should have tried several values of $M$. Explain why, in using your model to predict the population at year 2050, you selected the value of $M$ that you did.

**Exercise 8** In connection with Exercise 5:

a) Explain why the weight $w$ that a person can lift should behave like $w = k\,x^{2/3}$, where $x$ is the weight of the person.

b) Use this result to describe a fair method for doing away with weight "classes" altogether (often needed in small meets).

**Hint**: Assume that strength is roughly proportional to the cross-sectional area of muscle and that body weight is roughly proportional to body volume for persons of comparable fitness.

## BEFORE THE LAB

### Introduction

Simple harmonic motion is a fundamental model in science and engineering. The study of the simple differential equation. $y''(t) + k^2y(t) = 0$. is the first step in the mathematical analysis of the vibration of physical systems as well as in the analysis of the propagation of sound, light and electromagnetic waves. The elementary solutions to this differential equation, namely $\sin kt$ and $\cos kt$. are also used in modeling a host of other phenomena that have cyclical or near-cyclical behavior.

This project and the later one on *Fourier Frequency Decomposition* will show you how these important functions can combine in some interesting ways and, conversely, how "signals" (or functions) can be decomposed naturally into their frequency components.

**Exercise 1** Suppose the displacement $y(t)$ of a particle satisfies the equation, $y'' + 9y = 0$, and at time $t = 0$ has a 4 cm displacement and a 6 cm/sec velocity. Solve for $y(t)$ in both the forms $a \cos kt + b \sin kt$ and $A \cos k(t - \alpha)$.

**Exercise 2** For the function $f(x) = \cos x - \sin x$:

a) Write $f$ in the form $A \cos k(x - \alpha)$. **Note:** $\alpha$ turns out to be *negative*.

b) Find its range.

c) Find the largest domain including $x = 0$ for which $f$ is decreasing.

d) Find the inverse of $f$ and its domain and range.

## THE LAB

**Exercise 3** Using the HP, confirm that your work in Exercise 2 is correct. In particular, plot both $f$ and its inverse (restricted to the proper domains) on the same graph, thus demonstrating the theoretically predicted symmetry about the line $y = x$. **Note:** Since the HP does not plot $(x, y)$ in a 1-1 ratio. to see the symmetry more clearly, also simultaneously plot the third function, $y = x$.

**Exercise 4** This exercise gives the background for the study of sound and other traveling waves. Consider the function:

$$y = \sin Ax + \sin Bx.$$

where $A$ and $B$ are constants. We will study the effect when $A$ and $B$ change relative to each other. Note that $A$ and $B$ (the "circular frequencies") are measured in radian/sec. Set $A$ to a number of your choice, say between 4 and 10: this is your fundamental frequency and is to remain fixed.

a) First set $B = A$ and describe the predictable result for the frequency and amplitude of $y$.

b) Now change $B$ by a small amount, say 5 to 10%—*be sure to plot over a large enough time interval to see what is happening.* The results may surprise you—so carefully describe what you see regarding the amplitude and frequencies that you observe. For example, do you see the original frequency $A$, or has it been altered? Do you see a much lower frequency?

## AFTER THE LAB

**Exercise 5** You are to explain the two new frequencies that you observed in the second part of Exercise 4. **Hint**: The trigonometric identity,

$$\sin C \cos D = \frac{1}{2} \left( \sin(C + D) + \sin(C - D) \right),$$

will help. Compute the two new frequencies and point them out on your graph. In your experiment, what is the period of the "beats" (the beats occurs at the times of maximum amplitude)?

## BEFORE THE LAB

### Introduction

The project on *Simple Harmonic Motion* illustrated "harmonic motion" in which a mass (or a sound wave, etc.) vibrates at certain frequencies. These vibrations were represented by sine and cosine functions. Harmonic motion is one of the fundamental phenomena of nature, and is widely used by scientists and engineers in modeling these phenomena. In these studies, a key step is analyzing the "frequency content" of the vibrating object.

For example, the function

$$y = \sin 11x \cos x$$

can, by a trigonometric identity, be written:

$$y = \tfrac{1}{2}\sin 10x + \tfrac{1}{2}\sin 12x.$$

This latter form is often preferred because it reveals that function $y$ is the sum of the two components with circular frequencies 10 and 12 rad/sec (assuming $x$ is time in seconds). We say that we have "decomposed" the original function into its frequency components. Each component might represent a frequency of oscillation, or a color, or chemical compound or whatever the application calls for. The fruitful discovery that many functions can be so decomposed was discovered by J.B.J. Fourier in 1807. In particular, Fourier argued that most functions of interest can be written in terms of their frequencies; i.e. in the form:

$$\begin{aligned} f(x) &= b_1 \sin x + b_2 \sin 2x + \ldots + b_n \sin nx \\ &\quad + a_0 + a_1 \cos x + a_2 \cos 2x + \ldots + a_n \cos nx \end{aligned} \tag{11}$$

where $n$ is a integer. In principle, $n$ can be large or even infinite, but for our discussion, we will assume that $n$ is small. For simplicity, in our examples for this project, we will also assume that $f(x)$ is "odd" in which case all the cosine terms will be zero (as you will show in Exercise 1). The $a_k$ and $b_k$ in Equation 11 are called *Fourier coefficients* and in the first exercise, you will find out how to compute them for the case just described. We say that two functions $f$ and $g$ are *orthogonal* on $[-\pi, \pi]$, if

$$\int_{-\pi}^{\pi} f(x)g(x)\,dx = 0.$$

(The word "orthogonal" is used here since this definition really *is* a (vast) generalization of the notion of perpendicular lines in geometry. However, explaining these subtle connections would take us far afield of the subject examined in this project—this is a major topic in undergraduate Linear Algebra courses.)

**Exercise 1** Assume that $f(x)$ is an odd function (use the index of your text, if you don't know the definition of "odd function").

a) If $g$ is even, show that $f$ and $g$ are orthogonal on $[-\pi, \pi]$ (or, for that matter, on any interval $[-c, c]$).

b) Show that $\sin kx$ and $\cos px$ are orthogonal on $[-\pi, \pi]$ for any positive integers $k$ and $p$.

c) For $k$ and $p$ *distinct* positive integers. show that $\sin kx$ and $\sin px$ are also orthogonal on $[-\pi, \pi]$. (You will need a trigonometric identity here).

d) Suppose a given $f$ is to be decomposed as in (11) above. To find a coefficient $b_k$, for $1 \le k \le n$. multiply Equation 11 through by $\sin kx$ and integrate over $[-\pi, \pi]$. Use the orthogonality shown in parts (b) and (c) to show that

$$b_k = \frac{1}{\pi}\int_{-\pi}^{\pi} f(x)\sin kx\,dx.$$

e) Now similarly seek the $a_k$ in (11), showing that all the $a_k$ are zero.

**Exercise 2** Suppose you wish to expand the specific function, $f(x) = x$, in a Fourier series as in (11).

a) Show that $D(\sin kx - kx\cos kx) = k^2 x \sin kx$.

b) Using (a) and Exercise 1, show that

$$b_k = -2\frac{(-1)^k}{k}.$$

c) Explain why the $a_k = 0$.

We now show how the HP can be used to decompose a function into its frequencies or, stated differently, to find its Fourier expansion. Again we'll consider only the case when $f(x)$ is odd, hence only the sine terms in (11) above are needed.

We use only a finite number of sine terms, so the expansion will not be exact, but only an approximation to the given $f(x)$.

**Note**: Since all the functions we consider will be odd, in computing the $b_k$ coefficients, we integrate only over $[0, \pi]$ and then multiply the result by 2. (You be asked to justify this later).

```
@ FOU--Compute Fourier sine series
@ Required globals:
@      F--STOred expression for function to be expanded
@ Local variables on stack:
@      N--number of terms.
@ Remark: in the code "pi" stands for left-shift-SPC and "int"
@          stands for right-shift-COS
FOU:
<< -> N
   << 0                                  @ initialize sum to 0
      1 N FOR K                          @ set up FOR loop
         0 pi 'SIN(K*X)*F' EVAL 'X'      @ set up integration arguments
         int 2 * pi /                    @ integrate; multiply by 2/pi
         ->NUM                           @ convert to numerical mode
         'SIN(K*X)' EVAL *               @ multiply by sin(kx)
         +                               @ add term to sum
      NEXT
   >>
>>
```

We start with a cubic polynomial and a small $n$ to see what happens. If we first STOre the cubic polynomial, $x(\pi - x)(\pi + x)$, in F and the result of the 2 FOU in F2, and finally use the PLOT environment to plot F and F2 on the same graph we obtain a figure like Figure 18. **Tip:** Don't forget to speed up the numerical integrations by using the MODE environment to set your number display format to a low accuracy factor.

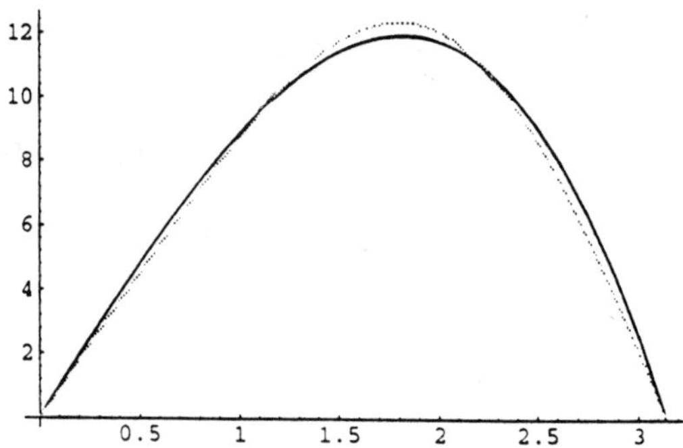


FIG. 18. The function $x(\pi - x)(\pi + x)$ and its two term Fourier approximation.

## IN THE LAB

**Exercise 3** For the above cubic, make $n$ larger, until the approximation $f_n$ agrees well with $f$ graphically. Hand in a graph with appropriate comments.

**Exercise 4** For $f(x) = x^2 \cot(x/2)$, start with a small $n$, and increase as necessary, compute a graphically accurate Fourier sine approximation $f_n$ on interval $[0, \pi]$. Print out one graph showing both $f$ and $f_n$.

Now consider the differential equation (or DE):

$$y'' + 8y = x^2 \cot(x/2) \tag{12}$$

The right side of Equation 12 is called the *forcing term*, because in mechanical applications it often represents an external force applied to the mass whose motion is described by $y$. Since this is a hard DE, we seek an approximate solution. From Exercise 4, you already have a good approximation to the difficult function $x^2 \cot(x/2)$ in the form:

$$b_1 \sin x + b_2 \sin 2x + \ldots + b_n \sin nx, \tag{13}$$

and you know an appropriate value for $n$. The idea is to seek a solution to the DE in (12) in the form:

$$y = c_1 \sin x + c_2 \sin 2x + \ldots + c_n \sin nx. \tag{14}$$

In Exercise 6 (after the lab) you will show that if (14) makes sense, then the $c$'s and $b$'s are related as follows:

$$c_k = \frac{b_k}{8 - k^2}, \qquad k = 1, 2, \ldots, n. \tag{15}$$

**Exercise 5** As just discussed, find the appropriate approximation (e.g. call it $y_n$) to the solution to (12) and plot it. Use the code in the on-line notebook to get the exact solution and its graph. Again using the notebook, plot the difference between the exact answer and your Fourier approximation. From your plots determine the size and location of the maximum error.

## AFTER THE LAB

**Exercise 6**

a) In Exercise 1d you found the formula for the $b_k$. Explain why if $f$ is odd, it is only necessary to integrate over $[0, \pi]$ as in the above HP code.

b) Explain the relationship in (15) by substituting (14) into the DE using (13) for the right side of the DE.

c) Your approximate solution to (12) found in doing Exercise 5 has the $n$ frequencies given in (13) and (14). However, from your earlier work with equations like $y''+8y = 0$, you might expect another frequency not accounted for. What is this frequency?

**Exercise 7** How do you tell if the "missing" frequency in the previous exercise is needed? Answer: Commonly in solving DEs like the one in (12), one has initial conditions that also need satisfying, that is, the values $y(0)$ and $y'(0)$ are specified in addition to the DE. If the approximate solution found in Exercise 5 does not satisfy these conditions, you need another piece. Think about how you solved "homogeneous" equations like $y'' + 8y = 0$ and see if you can get a solution for arbitrary initial conditions, $y(0) = a$ and $y'(0) = b$, where $a$ and $b$ are constants.

## BEFORE THE LAB

In this project we examine the mechanics of a rotating wheel which drives an arm. In practice, the arm could be connected to a piston or to a cutting blade. Mathematically, some interesting behavior takes place, including some rather unusual periodic functions and surprising results as the length of the arm approaches the radius of the wheel. Examine

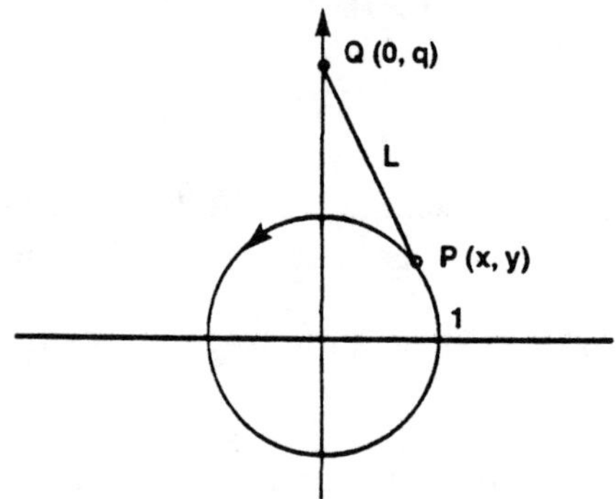


FIG. 19. Sketch for the piston problem.

Figure 19. The wheel is rotating in the counterclockwise direction at 2 radians/sec. The lower end of the arm at the point $P(x, y)$ is attached to the circle and rotates with it. The upper end of the arm at the point $Q(0, q)$ is attached to the vertical axis and can only move up and down on this axis.

**Exercise 1** In connection with the Figure:

a) How fast (linear velocity) is point $P$ moving?

b) Obtain this expression for $q$:

$$q(t) = \sin 2t + \sqrt{L^2 - \cos^2 2t}.$$

Hint: What are the coordinates of $P$?

c) Compute $q'(t)$.

**Tip:** Note that powers of trigonometric functions such as $\sin^2 x$ are coded as `(SIN(X))^2`.

**Exercise 2** You are not expected to know the "right" answer to the following questions—just use your imagination to gain an initial orientation to the situation. You will be able to later check your conjectures against what actually takes place.

a) Try to visualize the movement of point $Q$, as a function of $t$, as the wheel turns. First for $L$ about the size shown, then for $L$ slightly larger than the radius (1). Make rough sketches by hand.

b) Would you expect velocity of $Q$, $q'(t)$, to ever exceed that of $P$? Why?

c) For what location of $P$ would you expect $q'(t)$ to be largest?

## THE LAB

### Exercise 3

a) Compute $q'(t)$ and check against your hand work done earlier. **Tip**: Compute $q'$ with the $\partial$ key (review the *Newton's Method* project if necessary). If you want to see a simplified result, store the result in **EXPR** and use
[⇒] [SYMBOLIC] [Manip expr...] [COLCT] [NXT] [ok]
to simplify and then use
[⇒] [EXPR] [⇐] [VIEW] to see the simplification conveniently.

b) For $L$ considerably larger than the radius 1 (e.g. 2 or 3), graph $q(t)$ and $q'(t)$. Any surprises here? **Tip**: Use a [H-VIEW] of 0 to 9 and [V-VIEW] of -3 to 3.

c) Now experiment with $L$ only slightly larger than 1 obtaining graphs of $q$ and $q'$ that are rather strange (with nearly flat sections). Comment on what you see, especially try to explain these flat sections.

## AFTER THE LAB

### Exercise 4

a) Revisit your conjectures of Exercise 2 and compare with the actual results you obtained here.

b) A theoretical physicist is interested in what happens as $L \to 1$. In particular, she is interested in the maximum of $q'$ as $L \to 1$. Do you think, from your experiments, that this mysterious maximum value is finite or infinite? If finite, what do you think it is?

c) Briefly, if one were to explore, mathematically, the question raised in b), how would you pursue it?

## BEFORE THE LAB

### Integration of Even and Odd Functions

**Exercise 1** Evaluate the sequence of integrals, $I_n = \int_{-3}^{3} x^n\,dx$ for $n = 1, 3, 5$.

**Exercise 2** Suppose that $f$ is an *odd* function, that is: $f(-x) = -f(x)$. Prove that $\int_{-a}^{a} f(x)\,dx = 0$. **Hint**: Break the integral into two parts around $x = 0$ and use the *substitution* $u = -x$ on the integral from $-a$ to 0.

**Exercise 3** Suppose that $f$ is an *even* function, that is: $f(-x) = f(x)$. Prove that $\int_{-a}^{a} f(x)\,dx = 2\int_{0}^{a} f(x)\,dx$. This has practical use because evaluating at $x = 0$ is less error prone than evaluating at a negative number.

### The Method of Substitution

**Exercise 4** Let $I = \int \sin x \cos x\,dx$. use the substitution $u = \sin x$ to show that $I = \frac{1}{2}\sin^2 x + C$. Then use the substitution $u = \cos x$ to show that $I = -\frac{1}{2}\cos^2 x + C$. Is anything wrong here? Explain.

The function

$$n(x) = \frac{1}{\sqrt{2\pi\sigma^2}} e^{-(x-\mu)^2/2\sigma^2}$$

is the famous and important probability density for the *normal* distribution with mean value $\mu$ and standard deviations $\sigma$, see Figure 20.

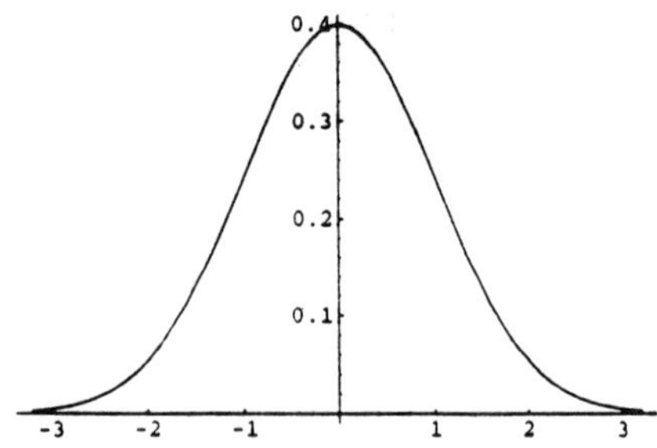


FIG. 20. The normal probability density.

If the "random variable" $x$ follows the normal distribution, then the probability that $x$ lies in the interval $[a, b]$ is defined as

$$P = \int_a^b n(x)\,dx.$$

It is often of interest to know the probability of lying within 1, 2, or in general, $k$, standard deviation of the mean. This is given by the integral

$$P(k) = \int_{\mu-k\sigma}^{\mu+k\sigma} n(x)\,dx. \qquad (16)$$

Indeed, everyone taking an elementary statistics course memorizes the following:

**Rule of Thumb for the Normal Distribution:** Two-thirds of the observed values lie within one standard deviation from the mean, 95% lie within two standard deviations, and 99% lie within three standard deviations.

Among other things, we'll be checking out this Rule of Thumb in this project. One famous (or infamous?) application is "grading on the curve": those students scoring within one standard deviation of the mean earn C's, while those scoring between one and two standard deviations above the mean earn B's and those scoring over two standard deviations above the mean earn A's and similarly at the low end for the D's and F's.

**Exercise 5** Perform the substitution $z = (x - \mu)/\sigma$ to show that

$$P(k) = \frac{1}{\sqrt{2\pi}} \int_{-k}^{k} e^{-z^2/2}\, dz. \tag{17}$$

Thus, $P(k)$ does not depend on $\mu$ or $\sigma$. This is important if you want to make tables for $P$—you need only one table instead of infinitely many. Notice also that this exercise is an application of substitution in integrals *not* aimed at getting an antiderivative.

As $k \to \infty$, we include all the probability, so you'd expect $\lim_{k\to\infty} P(k) = 1$ and an integration table will verify that this is so. What is the probability that a value lies within one standard deviation? After DEFining the function F(Z) as $e^{-Z^2/2}/\sqrt{2\pi}$, in response to the input:

'∫(-1,1,F(Z),Z)'

the HP gives 0.6827 (the exact answer depends on the display MODE set). Thus 68.3% of the probability is within one standard deviation of the mean. This is a more precise estimate than that given by the rule of thumb value (2/3 or 66.7%).

**Exercise 6** The *standard* normal distribution is the normal distribution with mean $\mu = 0$ and standard deviation $\sigma = 1$. Show that the probability density function for the standard normal is $p(x) = \frac{1}{\sqrt{2\pi}} e^{-x^2/2}$. Then show that the simplified form in Equation 17 of $P(k)$ is just the original form Equation 16 with $n$ replaced by $p$. (This explains why only the *standard* normal is tabled in handbooks.)

## THE LAB

**Exercise 7** **Plot** $\frac{1}{2}\sin^2 x$ and $-\frac{1}{2}\cos^2 x$ on the same graph and relate the result to Exercise 4. Note: In the HP, $\sin^2 x =$ `Sin[x]^2`.

**Exercise 8** Using the definition in Equation 17, show that $P(k) = \sqrt{2/\pi} \int_0^k e^{-z^2/2}\, dz$. Numerically evaluate $P(1)$, $P(2)$ and $P(3)$. Explain what these results mean in terms of probabilities and compare them to the rule of thumb given above. Explain how one can use Figure 20 to get a rough check that all is well.

**Exercise 9** For testing statistical hypotheses, it is also important to know the value of $k$ (not usually an integer) such that $P(k)$ equals a given value $v$. The values $v = .95$

("significant") and $v = .99$ ("highly significant") are especially popular. Find the $k$ values corresponding to the three $v$ values, .90, .95 and .99. **Hint**: Set up Newton's method, but watch out, the letters are nonstandard! In particular, $k$ is the independent variable, *not* a constant. Also, be careful about the starting value—a bad one could cause divergence.

## AFTER THE LAB

Nothing this time.

## BEFORE THE LAB

**Exercise 1** Do the following "by hand."

a) Find the Taylor polynomial of degree three, $P_3(x)$, by expanding the function $f(x) = e^x \sin x$ about $x = 0$. **A check:** You should get $f^{(iv)}(x) = -4e^x \sin x$ for the fourth derivative.

b) Using the remainder term in Taylor's theorem, determine an upper bound for the absolute value of the maximum error in $P_3(x)$ on the interval $[-1, 1]$.

**Exercise 2** The following idea of building on known expansions is often used to get an approximation for a function. Consider, for example, the function $g(x) = \sin(\sin x)$.

a) Explain why $\sin x \approx x - \frac{1}{6}x^3$ is a decent approximation for "small" $x$.

b) Suppose that the above small $x$ approximation is suitable for your purposes. You seek a similar approximation for $g(x)$. Let $u = x - \frac{1}{6}x^3$ (your approximation to $\sin x$), and now similarly approximate $\sin u$. Thus, by expanding in powers of $x$ obtain $g(x) \approx x - \frac{1}{3}x^3$. (You will see how well this works later in this project).

### Taylor Series with the HP

For finding a single Taylor polynomial around the origin, the HP provides a convenient interactive environment evoked by [⇒] [SYMBOLIC] [Taylor poly...] You will find that the **Taylor Polynomial** window is self-explanatory. For example, on (1) setting the **EXPR** field to `X*COS(2*X)`, (2) setting the **VAR** field to `X`, (3) the **ORDER** field to 5, and (4) pressing [ok] to **SYMBOLIC** evaluation, the result $X - \frac{12}{3!}X^3 + \frac{80}{5!}X^5$ appears on the stack (it takes some time). You can then press [EVAL] for a numeric result. The HP User's Guide (see page 20-13) describes the somewhat more involved procedure for computing the Taylor polynomial around an arbitrary point $x = a$.

Here is how to do the same operation directly from the keyboard:
'X*COS(2*X)' [ENTER] 'X' [ENTER] 5 [ENTER] TAYLR [ENTER]
Note the peculiar spelling of "Taylor" as **TAYLR**. We can write our own convenience Taylor procedure using our usual previously **STO**red expression for the function **F** of the variable **X**. We call it **MAC** (for MacLaurin), since it is limited to expansions around $x = 0$:

```
@ MAC--Implementation of MacLaurin series
@ Required globals:
@     F--stored expression of X for the function to be expanded.
@ Local variables:
@     N--order of expansion.
@ Limitations:
@     Only expands about X = 0

MAC:
<< -> N
   << F 'X' N TAYLR
   >>
>>
```

Repeat the above expansion using MAC.

Let's see how good our approximation is on the interval $[-1, 1]$. If we store the result in a variable and PLOT the approximation (called, say P5) and the function (called, say F) together, we get something like Figure 21 showing that the approximation is close, although it is breaking down at the ends of the interval. To better see the error, plot the difference

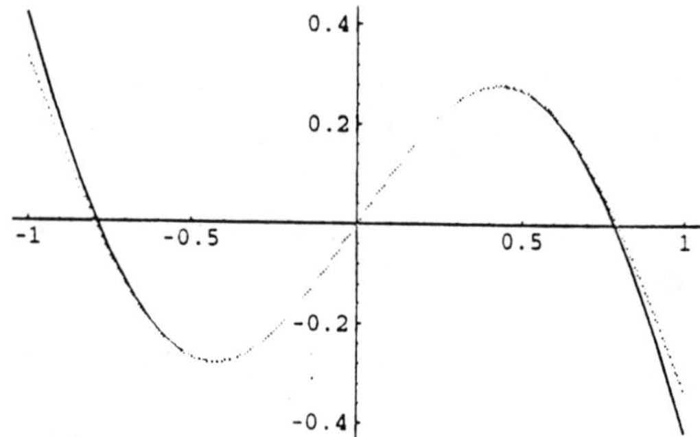


FIG. 21. $x \cos 2x$ and its $5^{\text{th}}$ degree Taylor polynomial at $x = 0$.

P5−F as shown in Figure 22. **Tip**: Enter F and P5, press [ − ] and store the result in DIFF. Then you'll have the difference painlessly stored and ready for PLOT. As usual, use [ (x,y) ] and the cursor keys to ascertain the scales on the resulting graph.

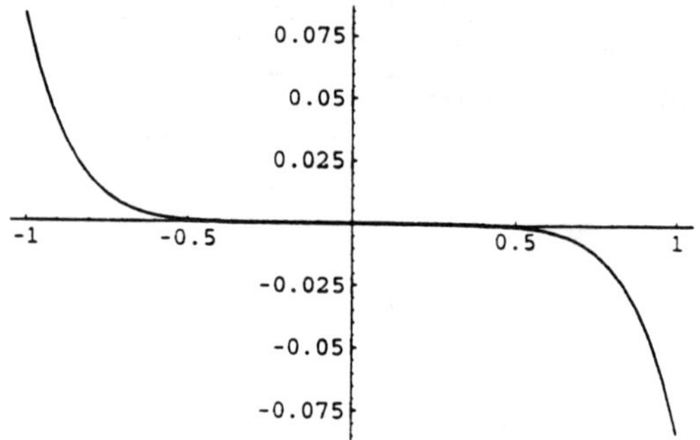


FIG. 22. The difference between $x \cos 2x$ and its $5^{\text{th}}$ degree Taylor polynomial at $x = 0$.

## THE LAB

**Exercise 3** For the function $f(x) = x \cos 2x$ discussed above:

a) Gradually increase $n$ until you have a good graphical approximation. Hand in a sketch of your final graph.

b) Suppose you wish to approximate $f(x)$ on the interval $[-1, 1]$ by an $n^{\text{th}}$ degree polynomial $P_n(x)$ to within an error of 0.005. By plotting the difference $f(x) - P_n(x)$, experimentally find the smallest $n$ that is adequate. Hand in a sketch of your final graph of $f(x) - P_n(x)$ showing the scales on the axes.

**Exercise 4** Following up on Exercises 1 and 2:

a) Use the MAC command to find the $P_3(x)$ in Exercise 1. Graph the error $f(x) - P_3(x)$ and comment on how the actual error compares with your upper bound in Exercise 1.

b) Use the MAC command to find a Taylor polynomial approximating $g(x) = \sin(\sin x)$, thus checking your earlier work in Exercise 2.

c) Plot $g(x)$ and your cubic approximation together on each of the intervals $[-0.5, 0.5]$ and $[-1, 1]$. Comment on the quality of your approximation in each case.

**Exercise 5**

a) Obtain series approximations to $\sin(\tan x)$ and $\tan(\sin x)$ about $x = 0$ and determine the first power of $x$ for which the series differ.

b) Since these two series look very much alike, you may think they represent essentially the same functions. Hand in plots of $\sin(\tan x)$ and $\tan(\sin x)$ on intervals $[0, \pi/4]$ and $[0, \pi]$ and comment.

c) One of the functions behaves wildly around $\pi/2$. Which one? Explain this strange behavior.

### Solving a Differential Equation via Taylor Approximations

Here, we use power series to solve a differential equation. On the interval $[-1, 1]$, we want to solve the differential equation,

$$y'' = g(x) = e^{-x^2} \sin x,$$

with initial conditions, $y(0) = 0.3$ and $y'(0) = -0.1$. This is reminiscent of the falling body problem, except now the right side, $g(x)$, is *not* the earth's gravitational constant. Thus, we will have trouble integrating $y''$ to get $y'$, etc. So we seek an approximation to the solution, $y$, as in the next exercise.

**Exercise 6**

a) Approximate $g(x)$ by a Taylor polynomial $P_n(x)$ for a large enough $n$ so that the approximation looks good on $[-1, 1]$.

b) Now integrate $P_n(x)$ twice, applying the initial conditions, thus getting a (hopefully) suitable approximation to our desired solution. Provide a graph of the resulting y.

### Using Series to Evaluate Limits

Since $\sin x = x - x^3/3! + x^5/5! + O(x^7)$, textbook authors easily construct "interesting" limit problems like

$$\lim_{x \to 0} \frac{\sin x - x + x^3/6}{x^5}.$$

With the series in hand, it is obvious that the quotient in the limit is $1/5! + O(x^2)$, so that the limit is $1/120$. Using the HP we could apply 7 `MAC` to the expression `'SIN(X)-X+X^3/6'` and obtain

```
'1/5!*X^5-1/7!*X^7'
```

then, simple hand simplifications reveal that

$$\frac{\sin x - x + x^3/6}{x^5} \approx \frac{1}{120} - \frac{x^2}{5040},$$

which not only makes it easy to read off the limit value, but also gives detailed information on the approach to the limit. **Tip**: The HP cannot get the Taylor polynomial for $(\sin x - x + x^3/6)/x^5$, which is why we expanded just the numerator in the example above.

You know, or will shortly know, that L'Hôpital's Rule gives a hand method for computing such limits—using our knowledge of series, it is easy to conjure up examples where L'Hôpital's Rule is fun to use or to conjure up examples where an exorbitant number of applications is required.

**Exercise 7** Use the series approach to evaluate the following limits:

a) $\lim_{x \to 0} \frac{\cos x - 1 + x^2/2}{x^4}$

b) $\lim_{x \to 0} \frac{\sin(x^3) - (\sin x)^3}{x^5}$

c) $\lim_{x \to 0} \frac{\sin(\tan x) - \tan(\sin x)}{x^7}$

**Tip**: Note that powers of trigonometric functions such as $\sin^2 x$ are coded as `(SIN(X))^2`.

## AFTER THE LAB

Repeat Exercise 7 using L'Hôpital's Rule.

## BEFORE THE LAB

Polar plotting is accessed by entering the PLOT environment ([⇒] [PLOT]) and [CHOOS]ing the TYPE as Polar. The interactive POLAR PLOT window is similar to those you have seen before, but there are a few special points to keep in mind:

1) Make sure that the ∠ field is set to RAD.

2) If you'd like to use $\theta$ as your independent variable, it is entered by pressing [$\alpha$] [⇒] F.

3) For polar plotting, you must distinguish between the horizontal plot limits (H-VIEW) and the plotting range (LO to HI). For example, if you are polar plotting

$$r = f(\theta) = \sin 2\theta,$$

then you know the maximum amplitude in any direction is 1. Thus the plot will fit in the window if you set both H-VIEW and V-VIEW to $[-1, 1]$. However, for plots that are more nearly in "one-to-one" ratio, we recommend setting H-VIEW to about about double V-VIEW. The default plotting range is LO= 0, HI= $2\pi$ and, as for this function, this is often adequate. However, in this project, you will need to change the upper limit, HI. To do this, press the [OPTS] menu item and change the HI field to the value you want, then press [OK] to return to the main plot window.

A plot of the polar function $r = f(\theta) = \sin 2\theta$ is shown in Figure 23—it is called a *four leaved rose*. To get a plot that looks like this on the HP, select H-VIEW= $[-2, 2]$, V-VIEW= $[-1, 1]$, and accept the default values LO= 0, HI= $2\pi$. **Tip:** To set HI to (say) $8\pi$, highlight the HI

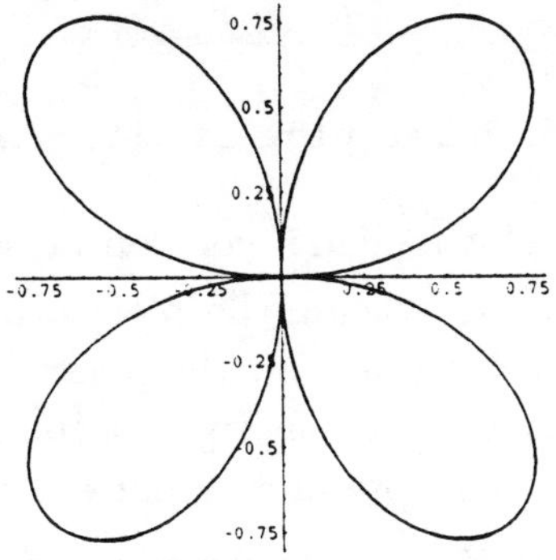


FIG. 23. Polar plot of $\sin 2\theta$ on $[0, 2\pi]$.

field and:
[NXT] [CALC]
The stack is then shown and you could press
[⇒] [$\pi$] 8 [×] [→NUM]
to compute the numerical value of $8\pi$. Then press [ok] once to return to the OPTS window where you will see the new HI value and a second time to return to the main plot window.

It is surprisingly difficult to predict the behavior of polar plots. For example, in the present example, since the period of $\sin 2\theta$ is $\pi$, we might think it sufficient to plot in the

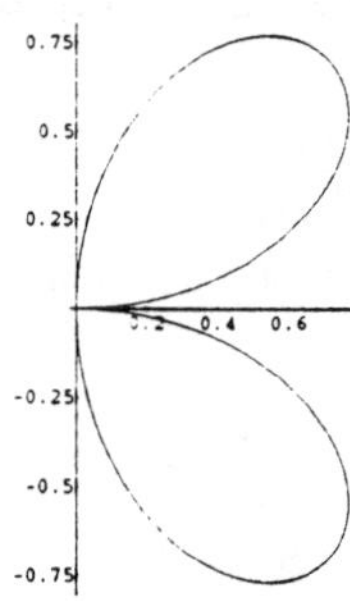


FIG. 24. Polar plot of $\sin 2\theta$ on $[0, \pi]$.

smaller range, 0 to $\pi$. But then we only get the right half of the graph, see Figure 24. Change `HI` to $\pi$ as in the above **Tip** and verify this. And if $\sin 2\theta$ produces a "four leaved rose," then surely $\sin 3\theta$ produces 6 leaves? No! It has three leaves and this time it is all done at $\pi$ instead of $2\pi$, see Figure 25. Change the function to $\sin 3\theta$ leaving `HI` as $\pi$ and

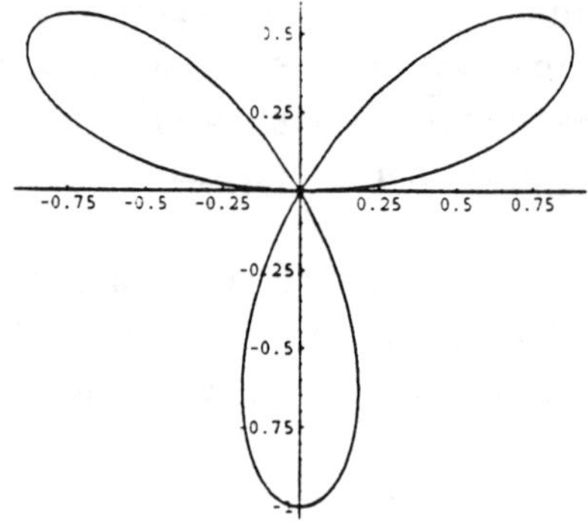


FIG. 25. Polar plot of $\sin 3\theta$ on $[0, \pi]$.

verify this. **Tip:** Once a polar plot is made, you can press the `TRACE` menu item and the coordinate viewer `(x,y)` to invoke tracing the plot with the `▷` key. This lets you see which $\theta$ value corresponds to which point on the polar graph. Trace the first leaf of the three leafed rose to verify that it ends at the origin at about 1.047 radians or 60° consistent with a traversal of the full three leaf rose in $\pi$ radians or 180° (3 times 1.047 is 3.141).

We hope you enjoy using the HP to explore and understand the nature of polar plots related to the ones we've just shown you—such a project would be out of bounds if we couldn't produce the plots automatically. Moreover, you should know that the "roses" come up in antenna design, earthquake radiation patterns, etc. So they are useful, as well as beautiful.

## THE LAB

**Exercise 1** Experiment with HP polar plotting on the equation $r = \sin k\theta$, $k = 1, 2, 3, \ldots$ and find out:

a) How the plot range needed to get a complete graph depends on the integer $k$.

b) How the number of "leaves" depends on the integer $k$.

**Remark:** Figures 23 and 25 provide correct information; your results should be consistent with them. Also, don't forget about the case $k = 1$!

**Exercise 2** For $r = \sin\left(\frac{\theta}{k}\right)$, with $k$ still a positive integer, find out how the plot range needed to get a complete graph depends on $k$.

**Exercise 3** (**Harder**) For $r = \sin\left(\frac{m\theta}{n}\right)$, with $m$ and $n$ positive integers, find out how the plot range needed to get a complete graph depends on $m$ and $n$. **Hints:** You can assume that $m$ and $n$ have no common divisors. You already have some evidence from Exercises 1 and 2. **Remark:** The full polar plot of something like $\sin\left(\frac{6\theta}{7}\right)$ is lovely, we hope you enjoy seeing it.

**Exercise 4** Describe what happens if we use an irrational factor $b$ in $r = \sin b\theta$. In particular, for $b = \sqrt{2}$ plot this function on each of the intervals $[0, 12\pi]$, $[0, 24\pi]$ and $[0, 48\pi]$.

**Exercise 5** Investigate the polar plots of the family of functions given by $r(\theta) = 1 + p\sin\theta$ for various values of the *positive* parameter $p$. Begin by trying the values $p = 3/4$, 1, and 4/3. Describe how the family looks in terms of the $p$ value. In particular, tell what happens as $p \to 0$ and as $p \to +\infty$.

**Exercise 6** Hand in the most interesting polar plot you can create. Think about using exponentials, powers of trig functions, etc. Just to get your juices going, try $r(t) = 3\cos^2 t - 1$ and r(t)=3 $\cos^3 t - 1$, but we're sure you can do a lot better!

## AFTER THE LAB

**Exercise 7** By consulting your text, put names on the plots you made in Exercise 5.

**Exercise 8** From your results in Exercise 5, describe how the two parameter family $r(\theta) = a + b\sin\theta$ looks for various values of the *positive* parameters $a$ and $b$.

## BEFORE THE LAB

In this project we demonstrate the ability of the HP to graph conics.

**Exercise 1** Classify as ellipses or hyperbolas and make a hand sketch for each of the following:

a) $\left(\frac{x-2}{4}\right)^2 + \left(\frac{y}{3}\right)^2 = 1$

b) $\left(\frac{x-2}{3}\right)^2 + \left(\frac{y}{4}\right)^2 = 1$

c) $\left(\frac{x-2}{4}\right)^2 - \left(\frac{y}{3}\right)^2 = 1$

d) $\left(\frac{y}{3}\right)^2 - \left(\frac{x-2}{4}\right)^2 = 1$

## THE LAB

To graph a conic enter the `PLOT` environment, highlight the `TYPE` field and `CHOOS`e `Conic`. The `CONIC PLOT` window is similar to ones you have seen before except that setting the `V-VIEW` is mandatory. **Tip:** You have to transpose the constant in filling out the `EQ` field. Also make adjustments to make the equation easier to type in. For example, you might enter the first equation as: `(X-2)^2/16 + Y^2/9 - 1`

**Exercise 2** Use the `Conic` plot `TYPE` to check and correct the hand sketches you made for Exercise 1.

**Exercise 3** Use the HP to plot and sketch each of the following conics. Use these plots and `(x,y)` to check the hand calculations needed to complete the square and so determine the coordinates of the center, the vertices, and the foci. **Tip:** Use a generous `H-VIEW` and `V-VIEW` to get a preliminary plot and then shrink these intervals according to what you see to get a nice plot.

a) $9x^2 - 16y^2 - 54x - 63 = 0$

b) $16x^2 + 25y^2 - 160x - 200y + 400 = 0$

c) $2y^2 - 3x^2 + 6x - 8y = 1$

d) $8x^2 + 4y^2 + 24x + 4y - 13 = 0$

e) $36x^2 + 36y^2 - 48x - 108y - 47 = 0$

**Exercise 4** Plot and sketch each conic below with the HP. Apply the $B^2 - 4AC$ test and explain why the result is compatible with the plot you obtained. Identify any degenerate cases. Warning: `XY` is *not* the same thing as `X*Y` with the HP!

a) $xy - 1 = 0$

b) $2x^2 - 4xy + 8y^2 + 7 = 0$

c) $3x^2 + xy + x - 4 = 0$

d) $3x^2 + 6xy + 3y^2 - x + y = 0$

e) $2x^2 + 6xy + y^2 + 8x + 4y + 4 = 0$

f) $x^2 - xy - 2y^2 + x - 2y = 0$

g) $x^2 + 2xy + y^2 - 2x - 2y + 1 = 0$

## AFTER THE LAB

Nothing this time.

## BEFORE THE LAB

Polar plotting is accessed by entering the PLOT environment (⇒ PLOT) and CHOOSing the TYPE as Parametric. The interactive PARAMETRIC PLOT window is similar to the other plot windows, but as with polar plotting, you must distinguish between the horizontal plot limits (H-VIEW) and the plotting range (LO to HI). Please read or re-read the discussions in the *Plotting in Polar Coordinates* project about setting H-VIEW, V-VIEW, and HI. The two functions $x(t)$ and $y(t)$ giving the parametric represention are entered between parenthesis in the EQ field.

For example, consider the equations $x = \cos t$, $y = \sin t$ with the parameter $t$ in the range $[-\pi, \pi]$, which give the familiar parametric representation of the unit circle (note that $x^2 + y^2 = 1$). Here are the HP settings to get a plot similar to that shown in Figure 26:
TYPE: Parametric ∠: Rad
EQ: '(COS(T),SIN(T))'
INDEP: T H-VIEW: −2 2
_AUTOSCALE: V-VIEW: −1 1
and in the OPTS menu set LO to 0 and HI to 6.28 (numerical $2\pi$).

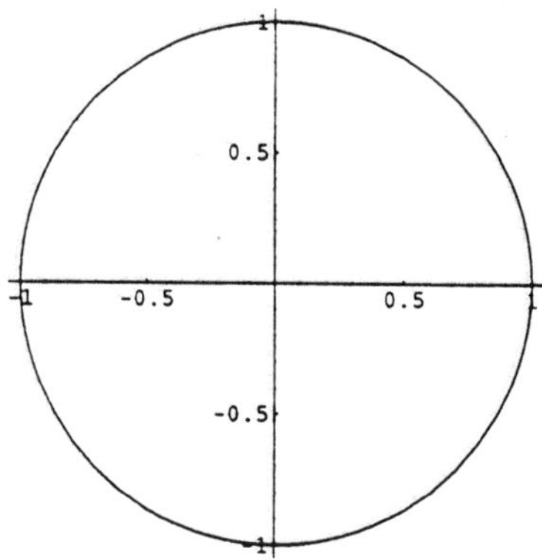


FIG. 26. A ParametricPlot of the unit circle.

## THE LAB

### The unit circle

The parameterization of the unit circle given above leads to a transversal of the circle with unit speed:

$$\begin{aligned} v(t) &= \sqrt{(x'(t))^2 + (y'(t))^2} \\ &= \sqrt{(-\sin t)^2 + (\cos t)^2} \\ &= 1. \end{aligned}$$

**Exercise 1** Show that the parametric equations $x(t) = (1 - t^2)/(1 + t^2)$, $y(t) = 2t/(1 + t^2)$ give an alternate representation of the unit circle (technically, the point $(-1, 0)$ is "missing"). Use TRACE, (x,y), and the ▷ key to trace along the curve to get the data to make a table showing the values of $x$, $y$ for the points $t = -5, -4, \ldots, 3, 4,$ and 5. Then compute the the speed $v$ and those values to your table. Describe in words how the speed varies as the point moves around the unit circle. Discuss the "missing point" in terms of limits. **Tip:** Be patient with the ▷ key, things get off to a slow start, so just keep it pressed for a while.

### The Hypocycloid

The *hypocycloid* is the curve generated by a circle of radius $b$ rolling on the inside of a larger circle of radius $a$, see Figure 27. The equations of the hypocycloid are given

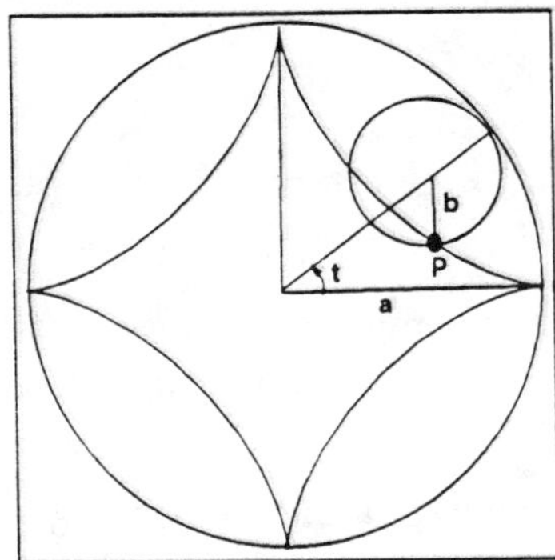


FIG. 27. A hypocycloid with $b/a = 1/4$.

parametrically as:

$$
\begin{aligned}
x(t) &= (a-b)\cos t + b\cos\frac{a-b}{b}t, \\
y(t) &= (a-b)\sin t - b\sin\frac{a-b}{b}t, \qquad a > b.
\end{aligned}
$$

We could factor out the scaling factor $a$ in each equation and study the family of hypocycloids in terms of the single parameter $b/a$. Alternately, we just set $a = 1$, but regard $b$ as representing the ratio $b/a$ and label our plots accordingly. With $a = 1$ and $b = 1/3$, the graph is shown in Figure 28.

**Exercise 2** Do graphical experiments to determine the period of the hypocycloid for the case when $a = 1$ and $b = 1/k$, with $k = 3, 4, 5, \ldots$. Write a sentence describing the nature of the graph.

**Exercise 3** Continuing the previous problem, describe what happens for $a = 1$, $b = 1/k$, as $k \to \infty$.

**Exercise 4** Determine the arclength of the limiting curve in the previous problem. **Hint:** most approaches go better if you get the arclength for one "lobe" and multiply by the number of lobes.

**Exercise 5** Do graphical experiments with the given hypocycloid code to determine its period when $a = 1$ and $b = j/16$, for $j = 1, 3, 5, \ldots, 15$. Write a paragraph describing what you observe about this sequence of plots.

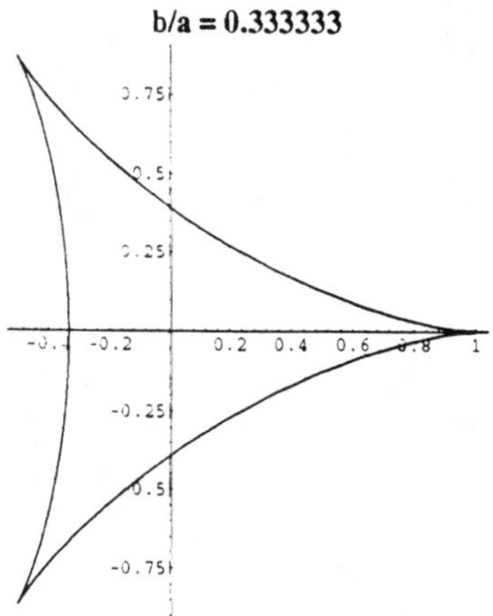


FIG. 28. A hypocycloid with $b/a = 1/3$. The plotting range is $[0, 2\pi]$.

**Exercise 6** Perform whatever experiments you find necessary to determine the period of the hypocycloid for $a = 1$ and $b = j/k$ where $j$ and $k$ are integers with $j < k$.

**Exercise 7** Hypocycloid pairs corresponding to $j/k$'s given by $1/3$ and $2/3$, or $1/4$ and $3/4$, or in general $1/k$ and $1 - 1/k$ have the same graph, but there is a difference, what is it? Be as quantitative as possible.

**Exercise 8** Investigate the special hypocycloid corresponding to $a = 1$ and $b = 1/2$.

### Another family of parametric curves

**Exercise 9** Study the family $x(t) = \cos^k t$, $y(t) = \sin^k t$, for $k = 1, 2, \ldots$. Describe the nature of the graphs and estimate the arclength of the limiting curve as $k \to \infty$. Can you explain your result intuitively?

## AFTER THE LAB

**Exercise 10** Look up (or derive yourself) the equations given above for the hypocycloid and give a clear and detailed exposition.

**Exercise 11** Verify the conclusion you drew in Exercise 6 directly from the parametric equations of the hypocycloid.

## BEFORE THE LAB

In this project, we return to our roots[2]. In our first serious project, *Graphical Equation Solving*, we solved cubic equations (i.e., found their roots) by plotting them and "zooming in" on the intersections with the $x$-axis. The central problem of that project was determining the normals to the parabola $y = x^2$ from a given point $(p, q)$. We are going to generalize that problem to three-dimensions, so please re-read the Illustrative Example and Exercises 4, 5, and 6 of the *Graphical Equation Solving* project. Exercises 7, and 10, of that project also are about the normal problem, but are of secondary importance to us now.

Your knowledge has increased greatly since you did that early lab project: you have mastered the derivative, the integral, Taylor Series and much more—you have a lot to be proud of! In particular, you now know all about Newton's Method for solving equations. In this project we will generalize Newton's Method from one equation in one unknown to two equations in two unknowns. So also please re-read the *Newton's Method* project, paying particular attention to the `Do Loop` implementation of this algorithm given there. Some texts give a careful derivation of Newton's Method in two dimensions—if yours does, you can just skim read the sketch of the derivation and interpretation given next and go on the to the *Illustrative Exercise*.

Recall the derivation of Newton's Method in *one* dimension. Given a function $y = f(x)$ and a starting value $x_0$ near a root (or zero) of the function (i.e., $f(x_0) \approx 0$), we seek a better approximation to the root. The idea of Newton's method is to replace the function by its linear approximation at the starting value, thus obtaining $y \approx f(x_0) + f'(x_0)(x - x_0)$. Then we put $y = 0$ and solve this linear equation for $x$ to get $x \approx x_0 - f(x_0)/f'(x_0)$. We take this approximate solution as our next starting point

$$x_1 = x_0 - f(x_0)/f'(x_0)$$

and repeat the process to obtain the general Newton iteration

$$x_{n+1} = x_n - f(x_n)/f'(x_n).$$

We know that if there really is a solution near $(p, q)$, Newton's method usually converges rapidly to that solution. (We also know that there are exceptional cases where Newton's method diverges!)

An alternate road to the Newton iteration result is to write the equation of the tangent line at $(x_0, f(x_0))$ as

$$y - f(x_0) = f'(x_0)(x - x_0)$$

and then solve for the point where the tangent line intersects the $x$-axis by putting $y = 0$ and denoting the intersection point by $x = x_1$.

Now consider the problem in *two* dimensions. Suppose we have the two equations

$$\begin{aligned} z &= f(x, y), \\ z &= g(x, y). \end{aligned}$$

[2]Most puns are awful and this one is no exception to the rule.

and a starting value $(p, q)$ near a simultaneous root of the equations (that is, both $f(p, q) \approx 0$ and $g(p, q) \approx 0$). Again, we seek a better value by using the linear approximation to the equations:

$$\begin{aligned} z &\approx f(x_0, y_0) + f_x(x_0, y_0)(x - x_0) + f_y(x_0, y_0)(y - y_0) \\ z &\approx g(x_0, y_0) + g_x(x_0, y_0)(x - x_0) + g_y(x_0, y_0)(y - y_0). \end{aligned} \tag{18}$$

Put $z = 0$ in each of these approximate equations and then solve for $(x, y)$. If we denote the solution by $(x_1, y_1)$, then we obtain the (usually) improved approximation:

$$\begin{aligned} x_1 &= x_0 - \frac{f(x_0, y_0)g_y(x_0, y_0) - g(x_0, y_0)f_y(x_0, y_0)}{f_x(x_0, y_0)g_y(x_0, y_0) - g_x(x_0, y_0)f_y(x_0, y_0)} \\ y_1 &= y_0 - \frac{g(x_0, y_0)f_x(x_0, y_0) - f(x_0, y_0)g_y(x_0, y_0)}{f_x(x_0, y_0)g_y(x_0, y_0) - g_x(x_0, y_0)f_y(x_0, y_0)} \end{aligned} \tag{19}$$

We'll give the HP code for this 2D iteration scheme below.

**Illustrative Exercise**: *Find the equation of the line through the point $P(4, 1, 0)$ that is normal to the paraboloid $z = x^2 + 4y^2$.*

**Solution:** We know that it is good scientific practice to solve the most *general* problems we can, so rather than doing algebra that has to be done all over again when the surface or the point is changed, we immediately generalize the stated problem to:

*Find the equation of the line through the point $P(p, q, r)$ that is normal to the surface $z = h(x, y)$.*

Introduce the point where the normal line intersects the surface as $(a, b, c) = (a, b, h(a, b))$. The parametric form of a line in 3-space is: $\boldsymbol{r} = \boldsymbol{r}_o + \boldsymbol{n}t$. Here, $\boldsymbol{r}_o$ denotes the given point, so that $\boldsymbol{r}_o = \{p, q, r\}$. Furthermore, $\boldsymbol{n}$ denotes the normal to the surface at the point $(a, b, h(a, b))$, so that $\boldsymbol{n} = \{h_x(a, b), h_y(a, b), -1\}$. Finally, $\boldsymbol{r}$ represents a vector to the general point on the normal. So, in particular, we can take it to be the point on the surface: $\boldsymbol{r} = \{a, b, h(a, b)\}$. With these replacements, the normal line equations give:

$$\begin{aligned} a &= p + h_x(a, b)t \\ b &= q + h_y(a, b)t \\ h(a, b) &= r - t \end{aligned}$$

We use the last equation to eliminate $t$, arriving at two equations for the two unknowns, $a$ and $b$:

$$\begin{aligned} a &= p + h_x(a, b)(r - h(a, b)) \\ b &= q + h_y(a, b)(r - h(a, b)) \end{aligned} \tag{20}$$

For the case of our paraboloid,

$$z = h(x, y) = x^2 + 4y^2,$$

Equations 20 reduce to (cf. Exercise 9):

$$\begin{aligned} 2a^3 + 8ab^2 + (1 - 2r)a &= p \\ 32b^3 + 8ba^2 + (1 - 8r)b &= q. \end{aligned} \tag{21}$$

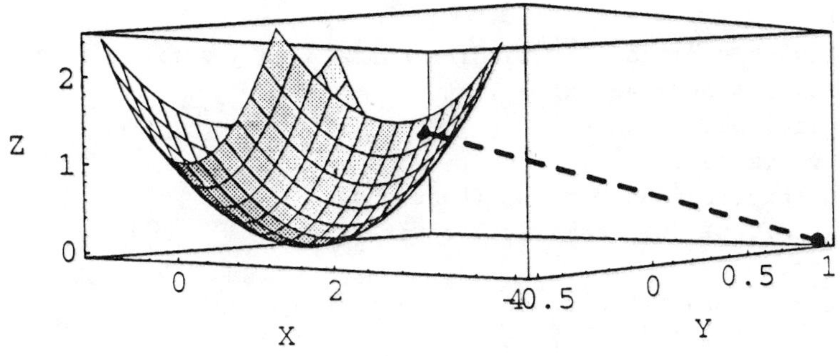


FIG. 29. Surface and normal for the Illustrative Exercise.

Now that we have the two nonlinear equations to solve for $a$ and $b$, and have an iteration scheme in hand, we come up against the really hard question: *how do we find a good starting point*? In the case of a single equation, we could often make a plot of $f(x)$ that revealed a good approximation of the root(s), but in two dimensions, directly plotting the surface $h(x, y)$ in a way that reveals a good starting value is usually too time consuming to be considered. Figure 29 does a good job for our special case, but the truth is, we need to know the answer in order to pick a good viewing direction and plot limits for $x$ and $y$. Getting a good three dimensional picture by doing an honest search is out of the question for most surfaces and points.

Here is a alternative *conceptual* approach—we are *not* suggesting that you actually carry it out: In place of three dimensional graphics, after filling in specific values for $p$, $q$, and $r$, we could try to make a sketch of each of the two dimensional curves in Equations 21. To do this, we could pick a $b$ in one of the equations, and solve it for the corresponding $a$-value(s). Then pick another $b$ and solve for $a$. Once we have enough points, we could plot that equation. Then we plot the other one the same way. The *intersections* of these two plots would give our starting point. Certainly a lot of work, but conceptually plausible because we could make use of our one dimensional plotting/Newton methodology in obtaining each $a$-value.

Since we understand how to proceed "in principle," in order to make this project practical to do, we will just give you plots of the two curves so that you can pick starting values from the intersections you see.

Figure 30 shows such a simultaneous plot for our case from which we can pick off the starting value $a = 1.1$, $b = 0.1$. Here is the code to carry out the Newton iteration, in this listing, the word "deriv" stands for the $\boxed{\partial}$ key.

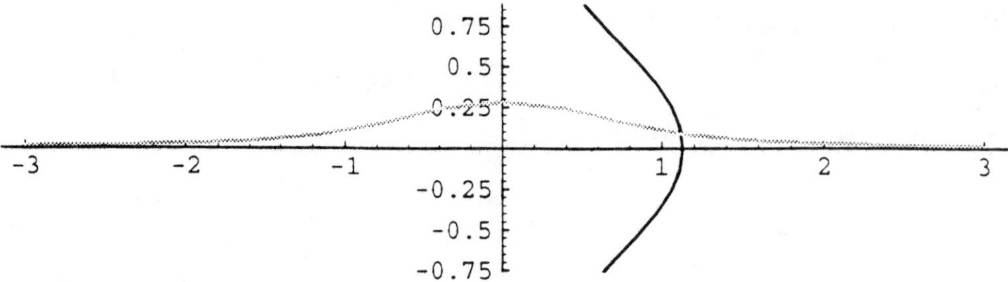


FIG. 30. Plot of $f = 0$ (dark) and $g = 0$ for the Illustrative Exercise.

```
@ NR2D--Newton's Method for finding normals to a surface from a given point
@ Puts iterates on stack as a list
@ Global variables:  None
@ Stored variables:
@     F,G--expressions for the functions.
@     DFX,DFY,DGX,DGY--numerical derivatives of F and G.
@ Local variables:
@     P,Q,R--coordinates of the given point.
@     A,B--initial guess at root.
@ Limitations:
@     Newton's method only converges locally in general
@     This is an application of Newton's Method in 2D, not the
@     general algorithm.
@ Note:
@     "deriv" stands for right-shift-SIN
<< -> P Q R A B                                 @ read from stack
   << '2*A^3+8*A*B^2+(1-2*R)*A-P' 'F' STO       @ define and store F
      '32*B^3+8*B*A^2+(1-8*R)*B-Q' 'G' STO      @ define and store G
      1 8 START                                 @ Do 8 Newton iterations
         F 'A' deriv 'DFX' STO                  @ numerical DF/DX
         F 'B' deriv 'DFY' STO                  @ numerical DF/DY
         G 'A' deriv 'DGX' STO                  @ numerical DG/DX
         G 'B' deriv 'DGY' STO                  @ numerical DG/DY
         A                                      @ a - num/denom in stack notation
         F DGY * G DFY * -                      @ num for a
         DFX DGY * DGX DFY * - /                @ denom and divide
         - EVAL 'A' STO                         @ subtract and update A
         B                                      @ b - num/denom in stack notation
         G DFY * F DGY * -                      @ num for b
         DFX DGY * DGX DFY * - /                @ denom and divide
         - EVAL 'B' STO                         @ subtract and update B
         A B 2 ->LIST                           @ put iterate on stack as list
      NEXT
   >>
>>
```

The result of running this code (with STD display mode) is:

```
1.12045115982     8.83389863455E-2
1.11997697384     8.86051111886E-2
1.11997671476     8.86052630684E-2
```

```
1.11997671476        8.86052630684E-2
1.11997671476        8.86052630684E-2
1.11997671476        8.86052630684E-2
1.11997671476        8.86052630684E-2
1.11997671476        8.86052630684E-2
```

Finally, we state our answer to the problem:
The normal line, $\mathbf{r} = \mathbf{r}_0 + \mathbf{n}t$. is given by

$$\begin{aligned} x &= p + h_x(a,b)t = p + 2at \\ y &= q + h_y(a,b)t = q + 8bt \\ z &= r - t \end{aligned}$$

where the given point is $\{p, q, r\} = \{4, 1, 0\}$, and $a \approx 1.11997671476$, $b \approx 0.0886052630684$.

**Exercise 1** Consider the problem of finding the point $x = a$. $y = b$ on the surface $z = h(x, y)$ that minimizes the distance between the surface and the given point $P(p, q, r)$. Show that $a$ and $b$ also satisfy Equations 20. State the meaning of this result.

## THE LAB

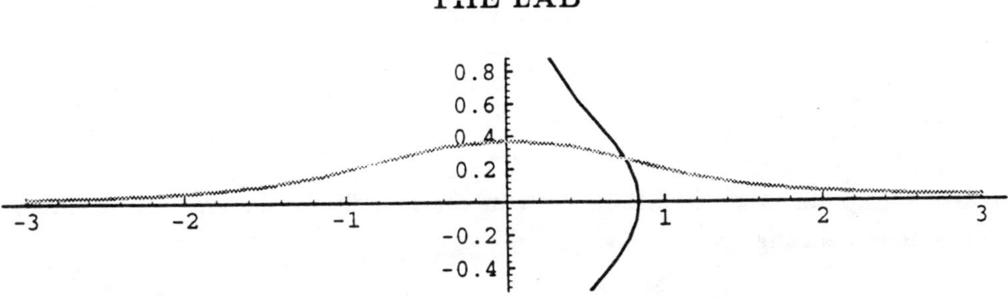


FIG. 31. Plot of $f = 0$ (dark) and $g = 0$ for Exercise 2.

**Exercise 2** Find the normal line to the paraboloid in the Illustrative Exercise, when the given point is $(2, 2, 0)$. Refer to Figure 31.

**Exercise 3** Find the minimum distance from the point $(2, 2, 0)$ of the previous Exercise to the paraboloid in the Illustrative Exercise. **Hint**: Given the result of Exercise 1, all you have to do is set up the formula for 3D distance and apply it appropriately.

**Exercise 4** Find the normal line to the paraboloid in the Illustrative Exercise, when the given point is $(4, 1, 4)$. Warning: There is more than one root in this case, find all the roots (and hence all the normals) that you can. Refer to Figure 32.

**Exercise 5** Which of the solutions in the last Exercise correspond to the *minimum* distance? What is the nature of the other solutions (i.e., are they local minima or something else?)

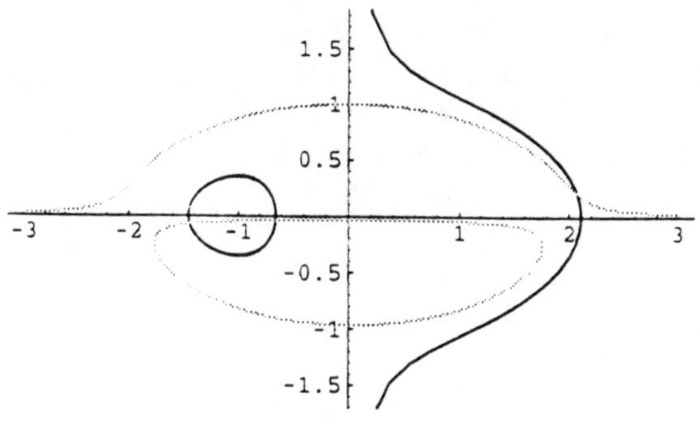


FIG. 32. Plot of $f = 0$ (dark) and $g = 0$ for Exercise 4.

## AFTER THE LAB

**Exercise 6** Derive the one dimensional Newton iteration for $x_1$ by using the tangent line approach.

**Exercise 7** Justify Equations 19 by solving the linear system

$$\begin{aligned} Ax + By &= r \\ Cx + Dy &= s \end{aligned}$$

and identifying the constants $A$, $B$. $C$, $D$, $r$, and $s$ in Equations 18 (after putting $z = 0$ in those equations).

**Exercise 8** Explain why Equations 19 can alternately be derived by finding the point $(x_1, y_1)$, where the tangent planes

$$\begin{aligned} z - f(p,q) &= f_x(p,q)(x-p) + f_y(p,q)(y-q) \\ z - g(p,q) &= g_x(p,q)(x-p) + g_y(p,q)(y-q) \end{aligned}$$

to the functions $f$ and $g$ at $(p,q)$ intersect the $z = 0$ plane.

**Exercise 9** Make the necessary identifications to specialize Equations 20 for the case of the paraboloid,

$$z = h(x,y) = x^2 + 4y^2,$$

and show that the non-linear equations to be solved in this case reduce to Equations 21.

## BEFORE THE LAB

In this project we'll show you how to do multiple integrals with the HP. As an application, we will compute the volumes and center of mass locations of some simple bodies in 2, 3, and 4 (!) dimensions.

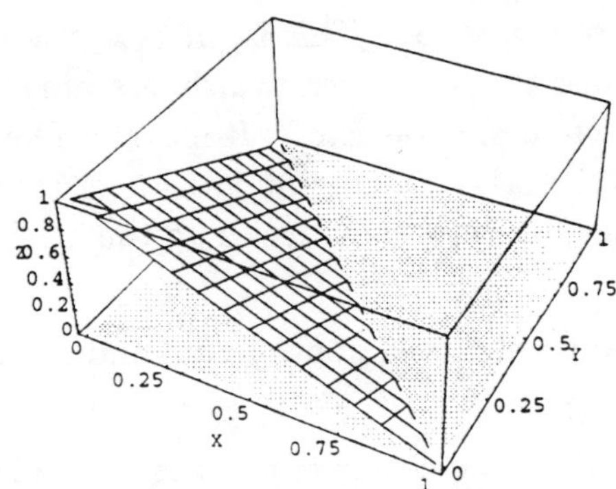


FIG. 33. Volume cut off by the plane $x + y + z = 1$ and the coordinate planes.

**Illustrative Example 1.** Find the volume of the region below the plane $x + y + z = a$ and satisfying: $0 \le x \le a$, $0 \le y \le a$, and $z \ge 0$.

**Solution:** The base of this region is the triangle in the $xy$-plane that is bounded by the coordinate axes and the line, $x + y = a$ (see Figure 33 in which we took $a = 1$). The mathematical expression for this volume is

$$\int_0^a \int_0^{a-x} \int_0^{a-x-y} 1 \, dz \, dy \, dx.$$

The HP cannot compute symbolic multiple integrals, so let's begin by setting $a = 1$ to make the integral numeric. To reduce the lengthy time required for a three dimensional integral, set the numeric mode to `Fix 3`. Here is the fast way to set the numeric mode:

[⇐] [MODES] [FMT] 3 [FIX]

Now use the equation builder to key in the multiple integral. Include all integral signs, use [SPC] to end limits and also to end the integrand . Each [SPC] after the integrand automatically supplies the lower case "d" for the differential ([▷] also ends limits, but it doesn't supply the "d"). With your previous experience with the HP, these hints probably suffice to let you enter the multiple integral. But just to be sure, we now give the keystrokes.

First start the equation builder:

[⇐] [EQUATION]

Then enter the first integral and limits:

[⇒] [∫] 0 [SPC] 1 [SPC]

continue with the second and third integral and limits:

[⇒] [∫] 0 [SPC] 1−X [SPC]

[⇒] [∫] 0 [SPC] 1−X−Y [SPC]

and finish with the integrand (just a 1 here) and the differentials:

1 [SPC] Z [SPC] Y [SPC] X [ENTER]

We remark again that the "d" in dZ, etc., is supplied automatically when you press SPC. You will now see the integral at level 1 of the stack expressed in algebraic notation. To evaluate it, simply press ⇐ →NUM and after some time passes, you'll get the numeric result 0.167. Surely, you suspect that the exact answer is a nice fraction! You can use the HP to check if there is such a nice nearby fraction like this:
⇐ SYMBOLIC NXT →Q
which gives the fractional result '1/6'. This is, in fact, the correct exact answer. While you are in this menu, also note the →Q$\pi$ item which lets you check for factors of $\pi$ (popular in integrals involving circles and spheres, etc.). **Remark:** The change of variables, $x = a\tilde{x}$, $y = a\tilde{y}$, $z = a\tilde{z}$ transforms the original integral with the $a$ into the one with $a = 1$ times a factor of $a^3$, so we conclude that the exact answer to the original integral is $a^3/6$.

**Exercise 1** Check the volume obtained above by a hand calculation.

**Illustrative Example 2.** Find the center of mass of the region in *Illustrative Example 1*.

**Solution:** We find the three components by integrating the components of the vector $\{x, y, z\}$ over the previously defined domain (i.e., replace the integrand 1 successively by $x$, $y$, and $z$) and dividing by the previously determined volume. With the extra spatial factor, the numerator now scales by $a^4$ (see the above **Remark**), so each component of the center of mass vector simply scales by $a$ itself and thus, we can again get the full answer by replacing $a$ by 1 and then restoring the scale factor at the end. It turns out that each of the three numerator integrals gives 0.042 with →Q producing the fraction '1/24' which again happens to be the exact answer. On dividing by the 1/6 in the denominator and restoring the $a$ factor, the center of mass is the vector $(a/4, a/4, a/4)$. **Tip:** Since the integrands are so similar, before evaluating, store the first one in a variable and later edit to get the next two. Also note that some components take significantly longer time than others on the HP.

**Exercise 2** Check the center of mass obtained above by a hand calculation. Explain why the components of the center of mass are equal.

## THE LAB

**Exercise 3** Use the HP to compute the volume and center of mass of the region *above* the plane $x + y + z = a$ and satisfying: $0 \leq x \leq a$, $0 \leq y \leq a$, and $0 \leq z \leq a$. (See Exercise 7.)

**Exercise 4** For several choices of $a$ and $b$, use the HP to compute the area and center of mass of the first quadrant of the ellipse $(x/a)^2 + (y/b)^2 = 1$. Use your results to conjecture the general results. Be sure to check your area conjecture for the case of a circle. **Tip:** Use →Q$\pi$. Also note that ▷ and *not* SPC is needed to terminate square roots in the equation builder environment.

**Exercise 5** Repeat the last problem for three dimensions, that is, compute the volume and center of mass of the portion of the ellipsoid $(x/a)^2 + (y/b)^2 + (z/c)^2 = 1$ that lies in the region $0 \leq x \leq a$, $0 \leq y \leq b$, and $0 \leq z \leq c$.

**Exercise 6** Repeat the last problem for *four* dimensions, that is, compute the hyper-volume and center of mass of the portion of the hyper-ellipsoid $(x/a)^2 + (y/b)^2 + (z/c)^2 + (w/d)^2 = 1$ that lies in the region $0 \leq x \leq a$, $0 \leq y \leq b$, $0 \leq z \leq c$, and $0 \leq w \leq d$. Use this result to figure out the hyper-volume of the entire hyper-ellipsoid and specialize the result to obtain the hyper-volume of the four-dimensional sphere of radius $r$. **Hint**: Surprisingly, the exact answer has the factor $\pi^2$ (*not* just a $\pi$), so look for the nearby nice (and again correct) exact answer by first dividing the HP decimal answer by $\pi$ and *then* using [→Qπ] on the result. Also, be aware that doing four dimensional integrals on the HP will take a long time. However, by this stage in your researches, you can probably conjecture the correct answers on the basis of one or at most two experiments.

## AFTER THE LAB

**Exercise 7** Show that the sum of your answer to Exercise 3 and the answer obtained in *Illustrative Example 1* is geometrically plausible.

## BEFORE THE LAB

### Work in filling a tank

Consider the work done in filling a tank with a fluid. For definiteness, let's assume that the fluid is originally in a shallow lake at $z = 0$ and that the tank is elevated with its $z$-dimension between $z = a$ and $z = b$, see Figure 34. For realistic values of $a$ and $b$,

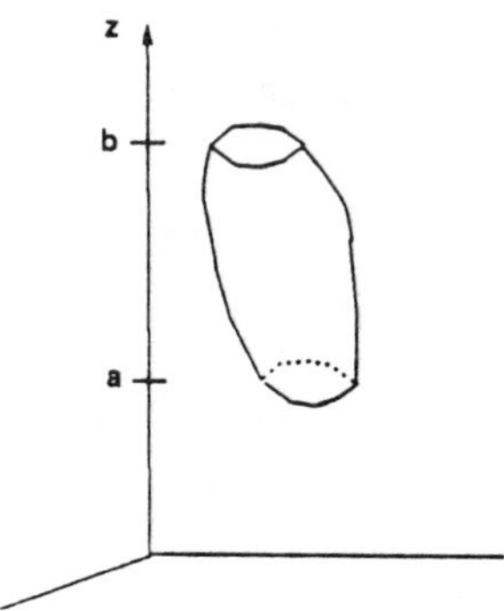


FIG. 34. Sketch of a body bounded between $z = a$ and $z = b$.

the constant force assumption is valid, that is, the force on a mass element $\Delta m$ is just its weight $\Delta m \cdot g$. Thus the work done in raising this mass element of fluid is $\Delta W = z \cdot \Delta m \cdot g$, where $z$ is the height to which this mass element is raised. Now $\Delta m = \rho \Delta x \Delta y \Delta z$, where $\rho$ is the density of the fluid, so we obtain

$$W = \rho g \iiint_V z \, dx \, dy \, dz,$$

where the subscript $V$ denotes integration over the volume of the tank. This is a familiar looking integral! We know that the $z$-component of the center of mass is given by

$$\bar{z} = \frac{\iiint_V z \, dx \, dy \, dz}{\iiint_V \, dx \, dy \, dz},$$

where the denominator is just the volume $V$ of the tank. Thus $W = \rho g V \bar{z}$ or $\boxed{W = Mg\bar{z}.}$ Here, $M$ denotes the total mass of the fluid filling the tank. The boxed result has two implications:

**theoretical implication** The work done in filling the tank is the same as the work in lifting a point particle of mass $M$ a distance $\bar{z}$ against gravity, i.e., to the height of the $z$-component of the mass center.

**practical implication** If we know the height $\bar{z}$ for the tank in question, and we know the total mass $M$ of the fluid that fills the tank, then computing the work is an easy calculation.

As an example of the second implication, consider the work done in filling a spherical tank of radius $R$ whose lowest point is at $z = h$. Obviously $\bar{z}$ is the height of the center of the sphere, so $\bar{z} = h+R$. Since the mass is $M = \rho \cdot V = \rho \cdot 4\pi R^3/3$, we get $W = 4\pi \rho g R^3(h+R)/3$ without doing any integrals!

**Exercise 1** In using the integral to evaluate the work that we just calculated directly by using the boxed equation, it is better to measure $z$ from the center of the sphere instead of from the ground. Show that in this case, the work integral becomes:

$$W = \rho g \int_{-R}^{R} \int_{\text{ymin}}^{\text{ymax}} \int_{\text{xmin}}^{\text{xmax}} (z + h + R)\, dx\, dy\, dz.$$

where the $y$ and $x$-limits remain to be figured out.

**Exercise 2** Use the method of cross sections to set up a one-dimensional integral for the work done in filling the spherical tank considered above (you are *not* required to evaluate this integral). Use the form of the work integral given in the previous problem.

**Exercise 3** Repeat the previous problem using spherical coordinates to derive a three-dimensional integral for the work.

**Exercise 4** Without doing any integrals, evaluate the work done in filling a vertical cylindrical tank resting on the ground with a fluid of mass density $\rho$. Assume that all the fluid is originally on the ground and that the cylinder has height $H$ and radius $R$.

**Exercise 5** Repeat the last problem for a horizontal cylinder.

### The inverse square law

The $F = -mg$ force law is just an approximation to the inverse square law that is valid near the surface of the earth. It is natural to wonder if the mass center trick also works for the full fledged inverse square law, which reads

$$F = -\frac{GMm}{r^2}.$$

Below, we shall pursue the question: *Is the force between two bodies the same as the force between two point masses separated by a distance equal to the distance of the mass centers of the original bodies?* The inverse square law force between two mass particles with respective masses $M$ and $m$ and separated by a distance $q$ has magnitude

$$F_{point} = -\frac{GMm}{q^2} \tag{22}$$

and acts along the line connecting the two particles.

## THE LAB

**Exercise 6** Use the HP to evaluate the integrals you set up in Exercises 1, 2 and 3 for the numerical values $h = 1$ and $R = 5$ and thus check the simple evaluation that we obtained from the boxed equation.

**Exercise 7** Use the boxed equation to help compute the work done in filling a tank in the first octant of the usual $x$-$y$-$z$ coordinate system that is formed by the plane $x + y + z = 1$ and the coordinate planes. Use the HP to get the needed volume and $z$ component of the mass center.

**Exercise 8** In this Exercise, we'll investigate the force between a particle of mass $m$ located at the point $(0, 0, q)$ on the $z$-axis and a thin square plate of negligible thickness $h$ and of density $\rho$ filling the region $-a/2 \leq x \leq a/2$, $-a/2 \leq y \leq a/2$, $-h/2 \leq z \leq h/2$. By symmetry, the resultant force is wholly in the $z$-direction. This vertical force between the particle and a volume element of the box is given by

$$\Delta F = \frac{Gm\,\Delta M \cos\alpha}{R^2},$$

where $R = \sqrt{x^2 + y^2 + q^2}$, $\Delta M = \rho h\,\Delta x\,\Delta y$, and $\cos\alpha = q/R$. Note that if the thin plate were collapsed onto its mass center (the origin), then the force between the true particle and the fictitious one at the mass center of the plate would be given by Equation 22, where $M$ is the total mass of the thin plate.

a) Draw a sketch and justify the formulas given above for $R$, $\Delta M$, and $\cos\alpha$.

b) Assuming that the vertical variation in force is negligible over the small thickness $h$, show that

$$F = -Gmh\rho q \int_{-a/2}^{a/2}\int_{-a/2}^{a/2} \frac{1}{R^3}\,dx\,dy.$$

c) To make our formula for the force look more like the point force formula, eliminate the density $\rho$ and show that the resulting expression is

$$F = -\frac{GMmq}{a^2}\int_{-a/2}^{a/2}\int_{-a/2}^{a/2} \frac{1}{R^3}\,dx\,dy.$$

d) Evaluate the integral in the previous part using the HP (you can set all constants to 1 including $q$ for now). Is $F = F_{point}$?

e) Take $a = 1$ and table the $F/F_{point}$ ratio for $q = 2^k$, $k = 0, 1, \ldots 8$. Does $F \to F_{point}$ as $q$ gets large? (Large $q$ results are referred to as "far field" results in the scientific literature.)

f) We know what expanding an expression for *small* $x$ means—get the Taylor series about $x = 0$. To get the series for *large* $q$, we can simply set $q = 1/x$ and get the series for small $x$. For example, to get the large $q$ series for $a/\sqrt{a^2 + q^2}$ replace $q$ by $1/x$ and simplify to get $ax/\sqrt{1 + (ax)^2}$. `STO`re `'A*X/√(1+(A*X)^2)'` in `F` and use the `MAC` procedure given in the *Approximation by Taylor Polynomials* project with input 3. The result is a bit messy, so simplify it with ⇐ SYMBOLIC COLCT. The result is $ax - (ax)^3/2$ or, in terms of $q$,

$$\frac{a}{q} - \frac{a^3}{2q^3}.$$

Expand the $F$ you obtained by integration for large $q$ and thus show that one gets $F \approx F_{point}$ in the far field. (We will say more about this "After the Lab.")

**Exercise 9** Repeat the previous problem for a thin circular disk of radius $a$. In evaluating the integral, use polar coordinates in the $xy$-plane. (Again, you can set all constants, except $q$ to 1).

## AFTER THE LAB

What does it really mean to say that a distance $q$ is "large" or is "small"? After all, suppose $q = 1\,\text{m}$. If we measure in microns, so that $q = 1,000,000\,\mu$, is $q$ now suddenly a "large" distance? Or if we measure in kilometers, is $q = 0.001\,\text{km}$ suddenly a "small" distance? Of course not! In real problems, there are natural scales and other quantities are "large" or "small" *relative* to these scales. In the inverse square law problems we posed above, the quantity $a$ provides the scale and large $q$ really means that the dimensionless quantity $q/a$ is large—or, equivalently, that $a/q$ is small.

## BEFORE THE LAB

**Illustrative Exercise** We now treat a problem of great historical interest: *Find the force between a sphere and a particle that is external to the sphere.* From our experience with such problems in the *Force Applications in 3D* project, we expect that in the "far field", the force will be nearly the same as the force between the original particle and a fictitious particle located at the center of the sphere. Also from that experience, we would not expect this replacement to be exact. But Newton showed that it *is* exact! We will repeat his calculation using the HP to help us. Along the way, we'll discover that in this problem, the HP needs a little help from us to simplify the result correctly.

First of all, we pick the $z$-axis to be the line from the center of the sphere to the particle. Thus we can write the location of the particle as $(0, 0, q)$. Denote the radius of the sphere by $a$ and its density by $\rho$. Since the particle is exterior to the sphere, we have $q > a$. With these choices, the geometry is as shown in Figure 35. By symmetry, the resultant force is

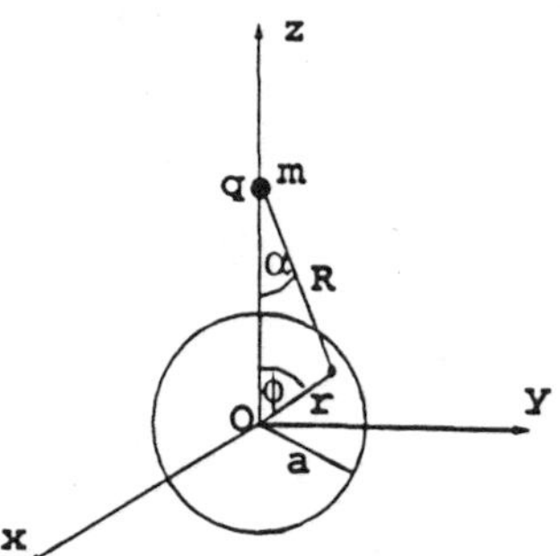


FIG. 35. Geometry of particle and sphere.

wholly in the $z$-direction. This vertical force between the particle and a volume element of the box located at the point with spherical coordinates, $(r, \phi, \theta)$, is given by

$$\Delta F = \frac{Gm\Delta M \cos\alpha}{R^2}.$$

Here, by the law of cosines, $R^2 = q^2 + r^2 - 2qr\cos\phi$. Also, we have $\Delta M = \rho r^2 \sin\phi \Delta r \Delta\phi \Delta\theta$, and $\cos\alpha = (q - z)/R$, with $z = r\cos\phi$. Thus, the gravitational attraction is given by

$$F = -Gm\rho \int_0^a \int_0^{2\pi} \int_0^{\pi} \frac{q - r\cos\phi}{R^3} r^2 \sin\phi \, d\phi \, d\theta \, dr.$$

Finally, to facilitate comparison with the force law between two points, we eliminate the density by using its definition: $\rho = M/V = 3M/(4\pi a^3)$ to obtain

$$F = -\frac{3GMm}{4\pi a^3} \int_0^a \int_0^{2\pi} \int_0^{\pi} \frac{q - r\cos\phi}{R^3} r^2 \sin\phi \, d\phi \, d\theta \, dr.$$

Ignoring the factor in front of the integral and using the numerical values $a = 1$ and $q = 2$, the HP gives 1.047 (after a significant pause). Using the →Qπ menu item gives $\pi/3$. Accounting for the constant in front of the integral, this agrees with the exact answer for any $a$ and $q$ which is (see the "After the Lab" section):

$$F = -\frac{GMm}{q^2}.$$

Hence, for the particle outside the sphere, we get *exactly* the same force as we would if we replaced the sphere by a particle at its center whose mass $M$ is the total mass of the sphere.

## THE LAB

**Exercise 1** Consider a hollowed out sphere, that is, the solid portion is the region $a < r < b$. Find the force between the hollowed out sphere and a particle outside this body. For numerical work, let $a = 1$, $b = 2$, and $q = 3$. Show that these evaluations illustrate that the inverse square law allows replacing the hollowed out sphere by a point at the center of mass.

**Exercise 2** Same as previous problem, but now the particle is located inside the hollow region. **Hint:** A very small numeric answer might be an approximation to __?

**Exercise 3** Consider a cylinder of radius $a$, height $2h$ and density $\rho$ filling the region $0 \leq x^2 + y^2 \leq a^2$, $-h \leq z \leq h$.

a) Derive a formula for the force $F$ between this cylinder and a particle of mass $m$ located at the point $(0, 0, q)$ on the $z$-axis. Write your result in terms of the mass $M$ of the cylinder. Then set $q = 2$ and the rest of the constants to 1 and evaluate the force using the HP.

b) Let $F_{point}$ denote the force between the particle and a fictitious particle of mass $M$ at the origin. Is $F = F_{point}$?

c) Take $a = 1$ and $h = 1$ and table the $F/F_{point}$ ratio for $q = 2^k$, $k = 0, 1, \ldots 8$. Does $F \to F_{point}$ as $q$ gets large (i.e., in the "far field")?

d) Expand the $F$ you obtained by integration for large $q$ and thus show that one gets $F \approx F_{point}$ in the far field.

**Exercise 4** Consider a particle of mass $m$ located at $(0, 0, q)$ on the $z$-axis and assume that a mass distribution of density $\rho$ is spread out over the entire $xy$-plane. Show that the attraction between the plane and the point is given by

$$F = -G\rho mq \int_{-\infty}^{\infty}\int_{-\infty}^{\infty} \frac{1}{R^3}\, dx\, dy,$$

where $R = \sqrt{x^2 + y^2 + q^2}$.

It can be shown that the force integral evaluates to $-2Gm\pi\rho$. **Remark**: This evaluation is too hard for the HP (this problem was designed for *Mathematica*). We stated it here, so you could be aware of this rather famous result. Accepting this result, comment on whether the inverse square law for extended bodies can always be replaced by point masses. Also comment on whether we can always make this replacement in the "far field."

## AFTER THE LAB

**Exercise 5** Do the integral in the *Illustrative Example* by hand. **Hint**: Use the law of cosines result for $R^2$ to replace the variable $\phi$ by the variable $R$.

**Exercise 6** Does the resulting force in Exercise 4 bear any resemblance to the force between two particles? Comment on the physical meaning of the result.

## BEFORE THE LAB

This project revisits the theme introduced in two earlier projects: *Work Along an Arc* and *Vectors and Work*. You now have enough mathematical tools to bring this subject to a logical conclusion. Until now the missing piece has been the notion of 'independence of path' for line integrals; that is, in going from point $A$ to point $B$ (in the plane or in $R^3$) does the path taken matter? If the application is arc length, for example, the obvious answer is 'yes'; and it comes as a bit of a surprise that in many line integral problems the answer is 'no'. The path independence issue is especially important in applications involving work, as one often wishes to minimize the amount of work, energy, etc. associated with a given task. In real life, the problem is usually compounded by other factors and trade offs are required. For example, if the required task is to move a load from point $A$ to point $B$ and the forces themselves are independent of the path taken, rarely does it *really* not matter about the path. Time is usually a factor, suggesting a straight line path. On the other hand, perhaps certain obstacles need to be honored, thus ruling out a straight line path. So the minimization problems suggested here, and elsewhere in your freshman-sophomore studies, are merely an introduction to those more realistic ones faced by modern science and industry. But the foundations are important, and we think that this foundation will be strengthened by this project.

We briefly review the situation in which a force vector $\boldsymbol{F} = (f, g)$ is applied to an object moving along a curve $C$ in the plane. The work done by the force is

$$\text{Work} = \int \boldsymbol{F} \cdot d\boldsymbol{r},$$

where $\boldsymbol{r} = (x, y)$ denotes the position on $C$. Two special cases are useful in the considerations to follow. If the curve $C$ is given by $y = y(x)$ for $a \leq x \leq b$, then we have

$$\text{Work} = \int_a^b \boldsymbol{F} \cdot (1, y'(x))\, dx = \int_a^b (f + g\, y')\, dx.$$

If, instead, one choses to parameterize $C$ with a parameter $t$ on $[0, 1]$ we have

$$\text{Work} = \int_0^1 (f\, x' + g\, y')\, dt$$

In most of the problems to follow you may use either representation, but since the first representation is just a special case of the latter (with $x(t) = t$), we give HP code for the parametric form:

```
@ WORK--2D Work integral
@ Global variables:
@      F,G--expressions in T for the components of the force vector.
@      X,Y--expressions in T for the components of the path vector.
@ Stored variables:
@      DX,DY--derivatives of X and Y.
@ Local variables:
@      A,B--limits of T integration.
@ Note:
@      "deriv" stands for right-shift-SIN
@      "int" stands for right-shift-COS
<< -> A B                                  @ read from stack
   << X 'T' deriv 'DX' STO                 @ DX/DT
      Y 'T' deriv 'DX' STO                 @ DY/DT
      A B 'F*DX+G*DY' EVAL 'T' int ->NUM   @ numerical work integral
   >>
>>
```

**Exercise 1** Suppose you are to power a boat from point $A = (0, 0)$ to point $B = (2, 1)$, and the primary consideration is the force of the wind which generally opposes you. You are to investigate the effect of taking different paths from $A$ to $B$. Note: the wind forces considered have negative components, indicating that the wind opposes the direction of the path. Hence the Work done by the wind is negative. But one may also think of the Work as that required to counteract the wind, thus producing a positive value. We prefer the latter point of view.

a) Suppose the wind force is $F = (-a, -b)$, for $a$ and $b$ positive. Compute the work done along the the straight line path between $A$ and $B$. Also compute the work done along a second path, of your choice. Does the path matter here? Why?

b) Now, due to the effect of the harbor you are entering, the wind force is $(-a, -ae^{-y})$. Again, compute the work along two paths as in part (a). Does the path matter here? Why?

c) Now let $F = (-a, -ae^{-y+cx})$ for $c > 0$. Compute the work along the straight line path and show that Work $= 2 + (1 - \exp(2c - 1))/(1 - 2c)$. (Note: for $c = 0$ you should agree with the Work in part (b)).

d) Now for a bit of a surprise. For the force of part (c), compute the work along the broken line path: up to (0,1) and over to (2,1). Explain why one does *not* have independence of path for this force. Comment on the fact that this path completely avoids the resistive part of the force represented by $e^{cx}$.

## THE LAB

An important type of problem in several fields of application is: Can we find the 'optimal path' (e.g., to minimize the work)? You are to explore this issue in regard to the wind force in part (c) of the previous exercise, $F = (-a, -ae^{-y+cx})$. Along with the paths you used in Exercise 1, you are to consider some new paths. Consider first, a family of parabolas described for $0 \leq t \leq 1$, by:

$$x = 2t, \quad y = 2t(\beta - 2\alpha t) \quad \text{where } \alpha > 0.$$

**Exercise 2** For the parabolic paths, and for the force $F = (-a, -ae^{-y+cx})$ with $c = 0.1$:

a) To go from (0,0) to (2,1), show that $\beta = (1 + 4\alpha)/2$.

b) Graphically show that for $\alpha$ 'large' the path swings 'north', i.e. it goes above the line $y = 1$ before ending at $y = 1$. (This could be bad). What's the situation for $\alpha = 0$? Moreover, find $\alpha$ such that the path comes into the final destination horizontally. (This could be good. You may want to find $\alpha$ analytically).

c) Now compute the Work in going from (0,0) to (2,1) along various parabolic paths. What is the situation regarding the minimizing of Work, when there is no restrictions on the parabolic path? Is this surprising in any way?

d) Same problem as in part (c), except now you have the restriction that the path cannot, due to the coast line, swing 'north'. There *is* an optimal path in this case; what is it?

e) We have so far assumed $\alpha > 0$. What happens for $\alpha < 0$, both graphically and regarding the Work?

**Exercise 3** Now consider a circuitous path from (0,0) to (1,2) described in the HP by:

```
x[t_] := 2t +  Sin[2Pi t]
y[t_] := t^2 + 1 -  Cos[2Pi t]
```

where $0 \leq t \leq 1$.

a) Compute the arc length of this path.

b) Compute the Work on this path for the current force. How does it compare with the Work along the parabolic paths?

c) Compute the Work on this path for the force $F = (-a, -ae^{-y})$. Compare with the results of Exercise 1b), and comment.

## AFTER THE LAB

**Exercise 4** Reflect on the work of the previous exercises.

a) What role did the parameter $c$ play in the amount of work required?

b) For $c > 0$, your experience should suggest that there *is* a parabolic optimal path in the case one cannot 'swing north'. What is it?

c) Is the path in (b) practical, say for navigating a boat?

d) What would you suggest as a practical path to (nearly) minimize the Work?

e) Recall that the work in Exercise 3(c) on the circuitous path was the same as that on the straight line path. How would you explain this strange result to a younger friend?

## BEFORE THE LAB

Here we will study the behavior of a swinging pendulum by numerically solving the determining (nonlinear) differential equation. Specifically, we look at the fundamental question: *how good is the linear approximation to a problem*? This is an important issue throughout science and engineering because typically the most accurate model we have for a problem is *nonlinear*. However, since nonlinear problems are often difficult to solve, it is tempting to make the linear approximation. In practice, it is easy to forget that the linear model, often studied in detail, is *not* the real thing and we can be led into erroneous conclusions. You will see in this study that, as is often the case, the linear model does pretty well when the key parameter is *small*, but can be totally inadequate as this parameter gets a bit larger. So a key question often is: "just what does *small* mean in this case?"

Figure 31 depicts a pendulum in which the bob has mass $m$ and the stem of length $L$ has negligible weight. Variable $y$ is the rotation angle, measured in radians, in the counterclockwise direction from the rest position. The differential equation for $y$ is:

$$mLy''(t) = -mg \sin y(t).$$

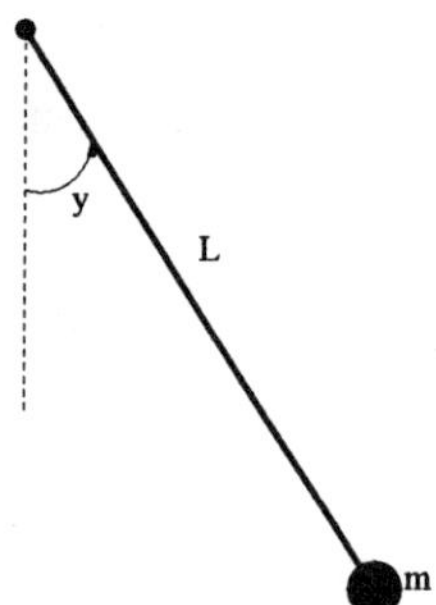


FIG. 31. The pendulum geometry.

**Exercise 1** Carefully derive this differential equation starting with the second law of motion: Force = mass × acceleration. **Hint**: notice that $Ly'(t)$ measures instantaneous linear velocity along a circle at time $t$. Think of the force acting on $m$, due to gravity, as a vector pointing downward, with two perpendicular components: one tangential to the circle of motion and the other 'radial' to the pendulum.

## IN THE LAB

For simplicity in the numerical studies, let's assume that $L = g$, so that the differential equation becomes

$$y''(t) + \sin y(t) = 0.$$

Here goes the key linearization argument. Most pendulums swing over a small angle $y$, hence $\sin y \approx y$. This leads to a linear differential equation which we suspect will adequately represent the action of the pendulum when $y(0)$ and $y'(0)$ are small:

$$y''(t) + y(t) = 0.$$

Now let's assume that at time $t = 0$ we displace the pendulum by amount $y(0)$ and gently release it (i.e., the initial velocity is zero). You can easily verify that the linear problem has the solution,

$$y(t) = y(0)\cos t.$$

We don't know the exact solution to the nonlinear problem; hence to compare the two solutions we need to solve the nonlinear equation numerically. Figure 37 shows a comparison extending over two periods of the linear solution when $y(0) = \pi/6$ (30 degrees).

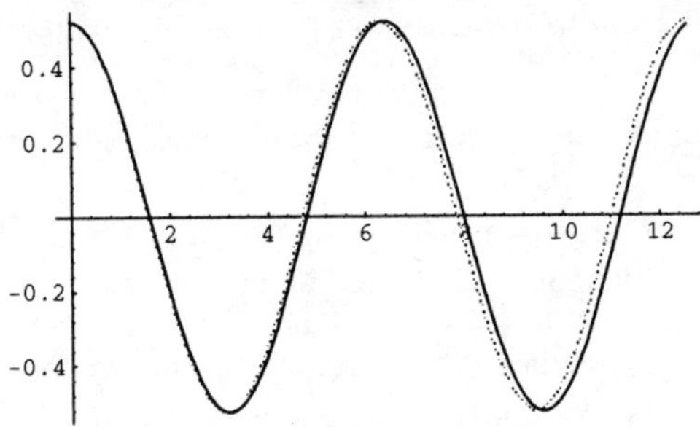


FIG. 37. Comparing the linear (light) and nonlinear (dark) pendulum when $y(0) = 30°$.

Chapter 19 of the HP User's Guide explains how to compute and plot solutions to differential equations. Since our system is second order, it needs to be rewritten in matrix form to accomodate the HP by introducing $v = y'$:

$$\begin{pmatrix} y \\ v \end{pmatrix}' = \begin{pmatrix} 0 & 1 \\ -\sin y & 0 \end{pmatrix} \begin{pmatrix} y \\ v \end{pmatrix}.$$

Furthermore, to fit the form used in Chapter 19, we rewrite this latter equation as

$$\begin{pmatrix} y \\ v \end{pmatrix}' = \begin{pmatrix} 0 & 1 \\ 0 & 0 \end{pmatrix} \begin{pmatrix} y \\ v \end{pmatrix} + \begin{pmatrix} 0 \\ -1 \end{pmatrix} \sin y.$$

On the HP, we represent the solution vector $(y \quad v)$ by the symbol `W`. To produce plots of the solutions with the HP, define `FUN` as `<<[ 1 0 ] W DOT SIN >>`, `V2` as `[0 -1]` and `M1` as `[ [0 1] [0 0] ]`. Then in the `PLOT` application

- Highlight the `TYPE` field and `CHOOS` `Diff Eq`.
- Make sure that angles are in radians.
- Set `F` to `'M1*W+V2+FUN'`.
- Set `INDEP` to `T`.
- Set `INIT` to 0.
- Set `FINAL` to 9.
- Set `SOLN` to `W`.
- Set `INIT` to the desired initial values using the `EDIT` menu item.

In the `OPTS` menu,

- Set `TOL` to .0001.
- Set `H-VIEW` to 0 10.
- Set `V-VIEW` to −.5 .5.

As usual you `OK` to the `PLOT` window and `ERASE` `DRAW` to get the plot. Note that as the project continues, you will have to reset some of the above fields.

**Exercise 2** Explore the *smallness* question in the linearization process.

a) Keeping $y'(0) = 0$, how large can $y(0)$ get before the solution to the linearized equation can be visually distinguished in the range $[0, 9]$?

b) Your textbook states somewhere that the pendulum of length $L$ has period $2\pi\sqrt{L/g}$. Is this really true or just an approximation? If not true, does the period change with $y(0)$? How? Investigate this by plotting the (nonlinear) pendulum's progress over several periods.

c) With the starting value $y(0) = \pi$ the two solutions are radically different (use an accurate approximation for $\pi$). Explain this on physical grounds.

**Exercise 3** Now set a zero initial displacement and study the behavior of the pendulum as $y'(0) = \omega_0$ varies. First observe that the solution to the *linear* pendulum problem is $y(t) = \omega_0 \sin t$. Thus, the linear approximation gives a solution with period $2\pi$ for all $\omega_0$.

a) To get an orientation, graph the solution for the nonlinear pendulum for the three values $\omega_0 = 1/4, 1/2, 3/4$ on the interval $[0, 9]$. Repeat for $\omega_0 = 1, 2, 3$ and $\omega_0 = 4, 5, 6$. On the basis of these results, give an overall description of how the solution varies with $\omega_0$ contrasting the nonlinear solution with the linear approximation.

b) You should have seen that the solution for $\omega_0 = 2$ is "special." Do plots near and on each side of this value and describe the behavior of the nonlinear solution for such values. What do you think is the physical meaning of the $\omega_0 = 2$ solution? (In terms of the quantities in the original equation, $\omega_0 = 2\sqrt{g/L}$.)

c) Compute the $\omega_0 = 2$ solution over a longer range, say $[0, 12]$. Do you believe what you see?

d) Repeat the previous part using a higher accuracy, say `TOL` set to .000001, to compute a more accurate answer.

## AFTER THE LAB

**Exercise 4** Discuss the results you got for the longer range with $\omega_0 = 2$ in the last problem. Are there any lessons to be learned about trusting "black boxes?" What is the role of the scientist in using automated solvers?

**Exercise 5** Interpret the large $\omega_0$ solutions physically—are they (nearly) periodic? Are the linear approximate solutions periodic? Do they make physical sense for large $\omega_0$?

**Exercise 6** Observe that $dy = y'\,dt$ and write the nonlinear equation in the differential form, $y'y''dt = -\sin y\;dy$. Integrate up to obtain

$$y'^2(t) - \omega_0^2 = 2[\cos y(t) - 1].$$

Use this result to derive an approximate expression for $y(t)$ when $\omega_0$ is very *large* (so that bounded quantities may be ignored). Is your analysis is in accord with the graphs you got earlier?

**Exercise 7** Use the result in the previous problem to give an intuitive explanation of the graphical results you got for $\omega_0 = 2$. **Hint**: What happens if $y(t_0) = \pi$ and $y'(t_0) = 0$ for some $t = t_0$? Look at the nonlinear differential equation!

**Exercise 8** An even better model of the pendulum would include things like friction at the pivot and air resistance. What effect would this have on our results?